Probability Theory and Examples

Probability Theory and Examples

Claire Montemar

ARCLER
PRESS

www.arclerpress.com

Probability Theory and Examples

Claire Montemar

Arcler Press
2010 Winston Park Drive,
2nd Floor
Oakville, ON L6H 5R7
Canada
www.arclerpress.com
Tel: 001-289-291-7705
001-905-616-2116
Fax: 001-289-291-7601
Email: orders@arclereducation.com

ISBN: 978-1-77361-315-4 (Hardcover)

ABOUT THE AUTHOR

Claire Montemar

Claire Montemar is a licensed Chemical Engineer with a real passion for teaching, learning and taking care of the environment. She obtained her degree from the University of the Philippines in 1999. She has worked in the manufacturing industry for a few years but spent most of her career as an academic tutor/instructor for elementary, high school and college students. Her forte is in the areas of mathematics and science, particularly algebra and chemistry. She currently works as a full-time freelancer, predominantly serving the needs of clients in the education industry.

TABLE OF CONTENTS

PREFACE

Probability is often a source of difficulty and anguish for students at both high school and undergraduate levels, because they don't know its importance and worth.

The foundation of probability was laid down by two French mathematicians of the seventeenth century – Blaise Pascal and Pierre De Fermat in connection with gambling problems. Later on it was developed by Jakob Bernoulli, Abraham De Mover, and Pierre Simon Laplace (1706-1749). The modern concept of probability theory which consists of stating few axioms and rules resulting from these axioms, was developed during the twenties and thirties of this century.

Probability theory has a wider pasture of application and is used to make sharp decisions in mathematics, management, operational research, sociology, astronomy, engineering, gambling, card games, psychology, physics and genetics where risk and ambiguity are involved.

Perhaps by providing a concise insight into where these concepts came from, how they were revealed, and how their uses in the past relate to how they are presently used and taught may offer students with the extra understanding they need to put these concepts to use, whether as measuring procedures or as functions. With this better insight, one would be able to see the value in studying Probability instead of an isolated unit from the subject.

This book aims to help the readers understand the amazing world of Probability, to enlighten them on how this subject have significance in real world problems.

Author

CHAPTER 1

HISTORY AND FUNDAMENTALS OF PROBABILITY

INTRODUCTION

Probability and statistics are fundamentally interrelated. Probability is often called the vehicle of statistics. The area inferential statistics in which we are mainly concerned with drawing inferences from experiments or situations involving an element of uncertainty, leans heavily upon probability theory. Uncertainty is also an inherent part of statistical inference. An inference is based on a sample, and a sample being a small part of the larger population, contains incomplete information. A similar type of uncertainty occurs when we toss a coin, draw a card or throw dice, etc. The uncertainty in all these cases is measured in terms of probability.

It is always clear what we mean when we make statements of the type that it is very likely to rain today nor I have a fair chance of passing the annual examination or will probably win a prize, etc. In each these statements, the natural state of uncertainty is expressed. But on the basis of past evidence, we have some degree of personal belief in the truth of each statement.

The foundation of probability were laid down by two French mathematicians of the seventeenth century – Blaise Pascal and Pierre De Fermat in connection with gambling problems. Later on it was developed by Jakob Bernoulli, Abraham De Mover, and Pierre Simon Laplace (1706-1749), The modern treatment of probability theory which consists of stating few axioms and rules resulting from these axioms, was developed during the twenties and thirties of this century.

Today, the probability theory has a wider field of application and is used to make intelligent decisions in economics, management operations, research, sociology, psychology, astronomy, physics engineering and genetics where risk and uncertainty are involved.

The word "probability" has two basic meanings: (i) a quantitative measure of uncertainty, and (ii) a measure of degree of belief in a particular statement or problem.

As the probability theory these days is best understood through the application of the modern set theory, therefore a brief description of the basic concepts of notation and operations of set theory that are relevant to probability, is given here.

An Aside Set

A *set* is any well-defined collection or list of distinct objects, e.g. a group of students, the books in a library, the integers between 1 and 100, all human beings on Earth, etc. The term *well defined* here means that any object must be classified as either belonging or not belonging to the set under consideration, and the term *distinct* implies that each object must appear only once. The objects that are in a set are called members or *elements* of that set. Sets are usually denoted by capital letters such as A, B, C, Y, etc. Elements are enclosed by braces to represent a set, e.g.

A = {a, b, x, y} or B = {1, 2, 3, 7}

If x is an element of a set A we write x A which is real as 'x belongs to A '' or x is in A.

If x does not belong to A, we write x A. The number of set A written as n (A) is defined as the number of elements in A.

A set that has no elements is called an **empty** or a **null set** and is denoted by the symbol It must be noted that {0} is not an empty set as it contains an element 0. If a set contains only one element it is called set a unit set or a singleton set. It is also important to note the difference between an element x and a unit set {x}. The elements of a set may be sets themselves.

A set may be specified in two ways. We may either give a list of all the elements of a set (the elements of a set (the Roster method), e.g.

A = {1, 3, 5, 7, 9, 11}; B = {a book a city a clock a teacher);

or we may state a rule that enables us to determine whether or not a given object is a member of the set (the rule method or set builder method), e.g.

A = {x is an odd number and x means that A is a set of all elements x such that x is an odd number and x is less than 12. The vertical line is read as "such that." The repetition or the order in which the elements of a set occurs, does not change the nature of the set. The number of elements present in it gives the size of a set. This number may be finite or infinite, thus a set is finite set.

The empty set is regarded as a finite set. Examples of a finite set:

(i) A = {1, 2, 3, ... , 99, 100}

(ii) B = {x $|x \text{ is a month of the year}$ };

(iii) C = {x $|x \text{ is a printing mistake in a book}$ }

(iv) D = {x $|x \text{ is a living citizen of pakistan}$ }; etc.

And the sets

(i) A = {x $|x \text{ is an even integer})$;

(ii) B = {x $|x \text{ is}$ a real number between 0 and 1 inclusive),

i.e., B = {x $|0 \leq x \leq 1\}$

(iii) C = $\{x|x \text{ is a point on a line}$ };

(iv) $D = x = \{x \; |x \text{ is a}$ sentence in the English language}; etc. are examples of an infinite set.

A set A is said to be in one-to-one correspondence with a set B when every element of set A can be made to correspond with the sequence of positive integers. A set is said to be non-denumerable when its elements cannot be enumerated.

Subsets

A set that consists of some elements or members of another set is called a subset of that set. For example, if B is a subset of A, then every member of set B is a also a member of set A and we write

B $\subset A$ or equivalently A $\supset B$

Which is read as B is a subset of A or is contained in A, or A contains B. For example, if

A = {1, 2, 3, 4, 5, 10} and B = {1,3,4}

When B $\subset A, i.e. B$ *is contained in A.*

It should be noted that set A is always regarded a subset of itself and an empty *set* $\varnothing$ *is considered to be (or accepted as) a subset of every set.* Two sets A and B are equal or identical, if and only if they contain exactly the same elements. That is, A = B if and A $\subset B$ *and* $B \subset A$. If set B contains some but not all of the elements of another set A while A contains each element of B, i.e. if

$$B \subset A \text{ and } B \neq A$$

then set B is defined to be a proper subset of A. The larger or the original set of which all the sets we talk about, are subsets is called the universal set or the space and is generally denoted by S or Ù. The universal set thus contain all possible elements under consideration. It is also regarded a subset of itself. A set S with n elements will produce 2^n subsets, including S and $\varnothing$.

Venn Diagram

A diagram that is understood to represent sets by circular regions, or their complements with respect to a rectangle representing the space S is called a Venn diagram, named after the English logician John Venn. Venn diagrams are used to represent sets and subsets in a pictorial way and to verify the relationship among sets and subsets. An example of a Venn diagram follows:

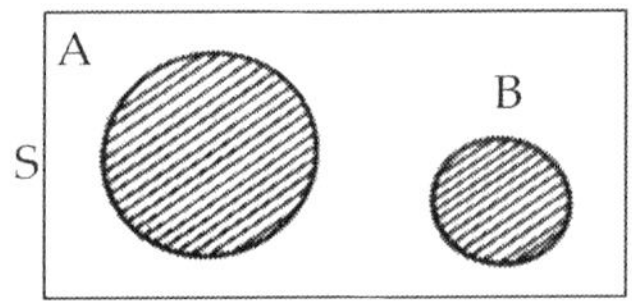

A Simple Venn Diagram

Operation on Sets

Let sets A and B be the subsets of some universal set S. Then these sets may be combined and operated on in various ways to form new sets which are also subsets of S. The basic operations are union, intersection, difference and complementation.

(i). The union or sum of two sets A and B denoted by A $\cup$ B, and read as *"A union B"* or *"A cup B"* means the set of all elements that belong to at least one of the sets A and B. That is,

$$A \cup B = \{x | x \in A \, or \, x \in B\}$$

By means of a Venn diagram, A $\cup B$ is shown in the areas below:

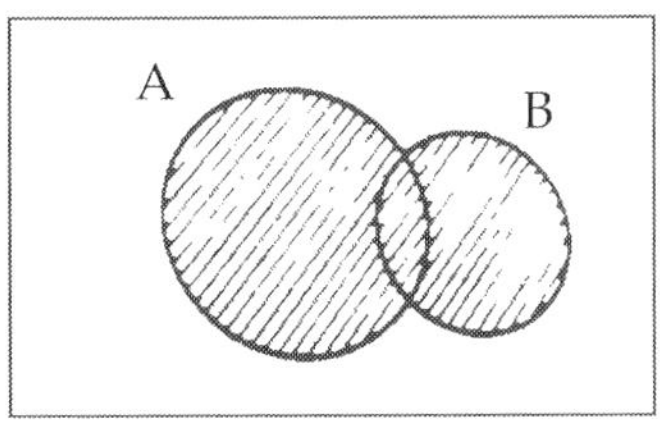

A U B is shaded

(ii). The intersection of two sets A and B, denoted by A $\cap B$ or by AB, and reads as *"A intersection B"* means the sets of all elements that belong to both A and B; that is

$$A \cap B = \{x = | x \in A \, and \, x \in B$$

Diagrammatically, A $\cap$B is shown in the area below :

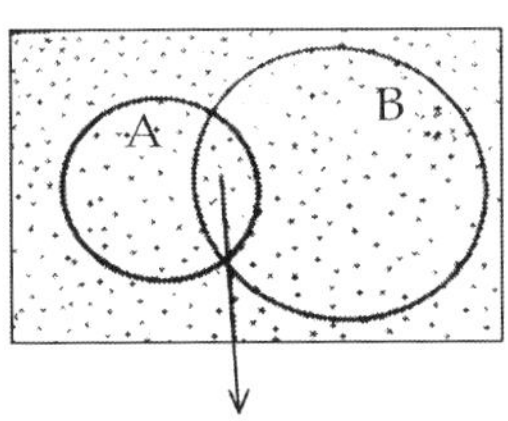

A∩B

The operation of union and intersection has been defined for two sets only. They may conveniently be extended to any finite number of sets.

Two sets A and B are defined to be disjoint or mutually exclusive or non-overlapping when they have no element in common, i.e., when their intersection is an empty set or $A \cap B = \varnothing$. On the other hand, two sets A and B are said to be conjoint when they have at least one element in common

(iii). The difference of two sets A and B, denoted by A – B or by A – (A $\cup B$).*is the set* of all elements of A which do not belong to B. Symbolically,

A-B = {x $|x \in B$ *and* $x \in$ *not* $B\}$

It is to be pointed out that in general A – B $\neq B - A$.

$B - A = \{x \; |x \in B$ *and* $\in$ *not* $A\}$.

The shaded area of the Venn diagram below shows the difference A - B.

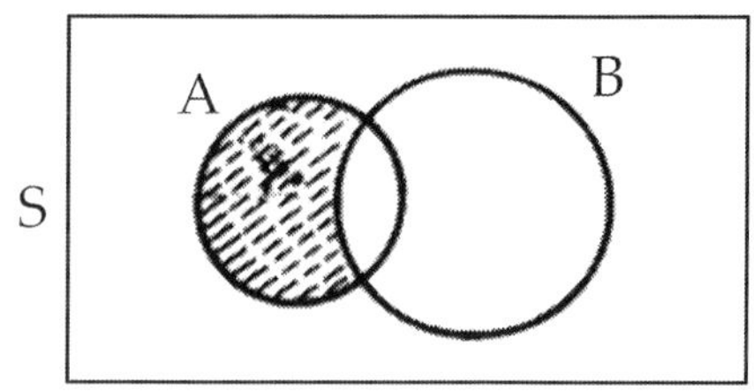

Difference A - B is shaded

It is noted that A – B and B are disjoint sets. If A and B are disjoint, then the difference A –B coincides with the set A.

(iv). *Complementation*. The particular difference S – A, that is, the set of all elements of S which do not belong to A, is called the complement of A and is denoted by $\overline{A}$ or by A^c

In symbols, $\overline{A}$ = {x $|x \in S$ *and* $x \in$ *not* $A\}$

The complement of S is the empty set $\varnothing$. The complement of A is shown by the shaded portion in the following Venn diagram

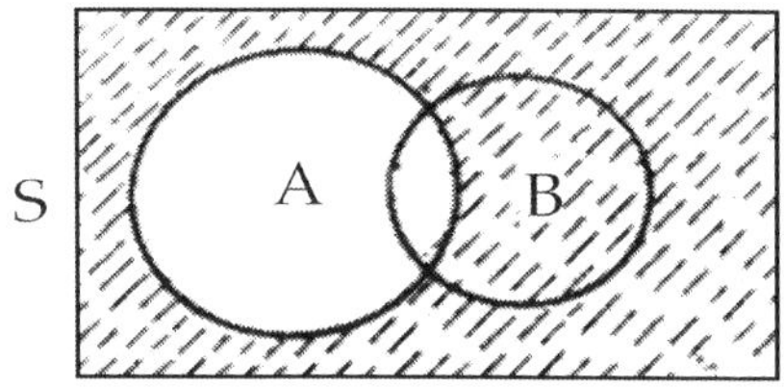

$\overline{A}$ is shaded

should be noted that A – B and $A \cap \overline{B}$, where $\overline{B}$ is the complement of set B, are the same set.

The Algebra of Sets

The algebra of sets provided us with laws which can be used to solve many problems in probability calculation. Let A, B and C be any subsets of the universal set S.

Then we have

(i) Commutative Laws

$A \cup B = B \cup A$ and $A \cap B = B \cap A$

(ii) Associative Laws

$(A \cup B \cup C = A \cup (B \cup C)$ and $(A \cap B) \cap C = A \cap (B \cap C)$

(iii) Distributive Laws

$(A \cap B) \cap C = (A \cap B) \cup (B \cap C) = (A \cup B) \cap (A \cup C)$

(iv) Idempotent Laws

$A \cup A = A$ and $A \cap A = A$

(v) Identity Laws

$A \cup S = S, A \cap S = A, A \cup \varnothing = A$, and $A \cap \varnothing = \varnothing$.

(vi) Complementation Laws

$A \cup A = S, A \cap A = \varnothing, (A) = A$, and $S = \varnothing, \varnothing = S$

(vii) De Morgan Law

$\overline{(A \cup B)} = \overline{A} \cap \overline{B}$, and $\overline{(A \cap B)} = \overline{A} \cup \overline{B}$

These laws can be visualized by means of the Venn diagrams.

Partition of Sets

A partition of set S is a sub-division of the set into non-empty subset that are disjoint and exhaustive, i.e. their union is the set S itself. This implies that

(i) $A_i \cup A_j = \varnothing, where i \varnothing j$;

(ii) $A_1 \cup A_2$

The subset in a partition are called cell. Let us consider a set S = {a, b, c, d, e}. Then {a, b} and {c, d, e} is a partition of S as each element of S belongs to exactly one cell.

Class of Sets

A set of sets is called a class, e. g., in a the set of the each lines is a set of points. This class of all subsets of set A is called m Power set of A and is denoted by p (a). For example, if A = {H, T}, then P(A) = { $\varnothing$,{H}, {T}, {H,T} }

Cartesians Product Sets

The Cartesian product of sets A and B denoted by A x B (read as "A cross B") is a set that contains all ordered pairs (x, y) where x belongs to A and y belonged to B. Symbolically, we write

A x B = {(x, y) $| x \in A and\ y \in B)$.

This set is also called the Cartesian set of A and B, named after the French mathematician Rene' Descartes (1596-1650). The product of a set by itself is denoted by A^2 . These concepts of products may be extended to any finite number of sets.

Let A = {H, T} and B = {1, 2, 3, 4, 5, 6}.

Then the Cartesian product set is the collection of the following twelve (2 x 6) ordered pairs:

$A \times B = \{(H,1);(H,2);(H,3);(H,4);(H,5);(H,6);(T,1);$

$(T,2);(T,3);(T,4);(T,5);(T,6)\}$

Clearly, these twelve elements together make up the universal set S when a coin and a die are tossed together. A die (plural, dice) is a cube of wood or ivory whose six faces are marked with dots as shown below where is the cube?;

It is relevant to note that, in general, $A \times B \neq B \times A$.

The product $A \times B$ may conveniently be found by means of the so-called tree diagram shown below;

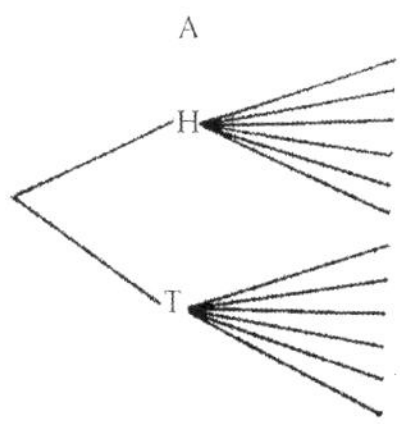

The "tree" is constructed from **left to right.** A tree **diagram** is a useful device for enumerating all the possible outcomes of two or more sequential events; the possible outcomes being represented by the individual paths or branches of the tree. This diagram is also used when we apply the multiplication rules to compute probabilities.

Relation and Function

A relation from a set A to set B is a subset of the Cartesian product of A x B. We usually call such a relation as binary relation. That is, a relation is an association between two or more objects. The set of the first elements of a binary relation is called the domain of the relation while the set of the second elements of the relation is called the range. For example, if a relation F = {(1, 4), (2, 7), (3, 12)}, then its domain is D = {1, 2, 3} and its range R = {4, 7, 12}.

A function is a rule that assign values in some manner from a set A to a set B that for each element x of A, there is a unique element y of B. Such an assignment is usually written as f: A B or *f is a function from A to B.* In other words, a function is special kind of binary relation that associates each element of the domain to element of the range. It is to be emphasized that a function is binary relation in which no first element is repeated. The value of the function f at x A is denoted by Y= f(x)

The variable x that represents the elements of the domain, is called the independent variable and the variable y = f (x) representing elements of the range is referred to as the dependent variable. Such functions are also called single-valued functions. A function, whose range consists of numbers, is called a numerical function. A function whose domain and range consists of sets of real numbers, is said to be a real-valued function of a real variable. A function f(x) is defined to be an even function if for every x in a certain range f(-x) = f(x), e.g. f(x) = functin f(x) having the property such that f(-x) = -f(x), is said to be an odd function, e, g., f(x) =

Random Experiment

The term "experiment" means a planned activity or process whose result yield a set of data. A single performance of an experiment is called a trial. The result obtained from an experiment or a trial is called an outcome.

An experiment, which produces different results even though it is repeated a large number of times under essentially similar conditions is called a random experiment. balanced die. Drawing of a card from a well-shuffled deck of 52 playing cards. A random experiment has three properties:

(i) It can be repeated, practically or theoretically, any number of times;

(ii) It always has two or more possible outcomes. An experiment that has only one possible outcome is not a random experiment;

(iii) The outcome of each repetition is unpredictable, i.e., it has some degree of uncertainty.

It is to be remembered that an ordinary deck of playing cards contain 52 cards arranged in 4 suits of 13 each. The four suits are called diamond, heart, club and spades; the first two are red and the last two are black. The face values are called denominations, of the 13 cards in each refer to denominations ace, 10 jack, queen and king. The face cards are king, queen and jack. These cards are used in various games such as whist, bridge, poker, etc.

Sample Space

A set consisting of all possible outcomes that can result from a random experiment (real or conceptual), is defined to be a sample for the experiment (and is denoted by the letter S). Each possible outcome is a member of the sample space and is called a sample point in that space. For instance, the experiment of tossing a coin results in either of the two possible outcomes: a head (H) or a tail (T); landing on its edge or rolling away is not considered. The sample space for this experiment may be expressed in set notation as S = {H, T}. The sample space for tossing two coins once (or tossing a coin twice) will contain four possible outcomes denoted by S = {HH, HT, TH, TT}. Clearly, S is the Cartesian product $A \times A$, where $A = \{H, T\}$. Similarly, the sample space S for the random experiment of throwing two six-sided dice can be described by the Cartesian product A x A, where A = {1, 2, 3, 4, 5, 6}. In other words,

$$S = A \times A = \{(x, y) | x \in A \, and \, Y \in A\}$$

where x denotes the number of dots on the upper face of the first die, and y denotes the number of dots on the upper face of the second die; and S contains 36 outcomes or sample points, which may also be conveniently represented in the following manner:

$$\begin{aligned} S = \{&(1,1),(1,2),(1,3),(1,4),(1,5),(1,6) \\ &(2,1),(2,2),(2,3),(2,4),(2,5),(2,6) \\ &(3,1),(3,2),(3,3),(3,4),(3,5),(3,6) \\ &(4,1),(4,2),(4,3),(4,4),(4,5),(4,6) \\ &(5,1),(5,2),(5,3),(5,4),(5,5),(5,6) \\ &(6,1),(6,2),(6,3),(6,4),(6,5),(6,6)\} \end{aligned}$$

This sample space may more briefly be expressed as

$$S = \{(i, j) | i = 1,2,3,4,5,6 \; ; j = 1,23,4,5,6)$$

A sample space that contains a finite number of sample points is said to be a finite sample space. It is defined to be a discrete sample space if the sample points can be placed in a one-to–one correspondence with the positive integers or if it is finite. If it satisfies neither of these criteria, it is called continuous. It should be noted that our sample space S would be finite unless otherwise stated.

Events

An event is an individual outcome or any number of outcomes (sample points) of a random experiment or a trial. In set terminology, any subset of a sample space S of the experiment is called an event. An event that contains exactly one sample point is defined as a simple event. A compound event contains more than one sample point and is produced by the union of sample events. For instance, the occurrence of a sum of 10 with a pair of dice, is a compound event, as it can be decomposed into three simple events (4, 6), (5, 5) and (6, 4).

An event A is said to occur if and only if the outcome of the experiment corresponds to some element of A. The event "not-A" is denoted by A or and is called the negation (or complementary event) of A. For example, the complement of "heads" is "tails" for tossing of one coin, the complement of at least one head on 4 tosses of a coin is no head. A sample space consisting of n sample points can produce different subset or simple and compound events as each sample point can either be included or excluded in forming a subset.

To illustrate, let us consider a sample space S containing three sample points, i.e. S = {a, b, c}

Then the eight possible subsets are

$$\varnothing,\{a\},\{b\},\{c\},\{a,b\},\{a,c\},\{b,c\},\{a,b,c\}$$

Each of these subsets is an event. The subset {a, b, c} is the sample space itself and is an event. It always occurs and is known as the certain event. The empty set $\varnothing$ also an event, sometimes known as impossible event. Because it can never occur.

This case of $2^3 = 8$ subsets (events) can be thought of as a field which is denoted by j.

These events have the following characteristics:

(i) The union of any number of events will result in a set that belongs to j.

(ii) The intersection of any number of events will result in a set that belong to j.

(iii) The difference of any two events belongs to j.

(iv). The complement of any event belongs to j.

Mutually Exclusive Events

Two events A and B of a single experiment are said to be mutually exclusive or disjoint and only if they cannot both occur at the same time. That is, they have no points in common. For instance, when we toss a coin, we get either a head or a tail, but not both, and the two events head and tail are therefore mutually exclusive; when a die is rolled, the outcomes are mutually exclusive as we get one and only one of six possible outcomes 1, 2, 3, 4, 5 or 6. Similarly, a student either qualifies or fails, a single birth must be either a boy or a girl, it cannot be both, etc. Three or more events originating from the same experiment are mutually exclusive if pairwise they are mutually exclusive, e.g. if we draw a card from an ordinary deck of 52 playing cards, it can be both a king and diamond. Therefore, kings and diamonds are not mutually exclusive. Similarly, inflation and recession are not mutually exclusive events.

If a sample pace $S=\{1,3,5,7,9\}$ and an event $A=\{1,3,5\}$ and another event $B=\{9\}$, then A and B are disjoint sets and they are said to be mutually exclusive events. In tossing a coin, the sample space $S=\{H,T\}$. Now, if event $A=\{H\}$ and event $B=\{T\}$, then $A\, and\, B$ are mutually exclusive events.

Exhaustive Events

Events are said to be collectively exhaustive, when the union of mutually exclusive events is the entire sample space S. Thus, in our coin-tossing experiment, head and tail are collectively exhaustive set of events. A group of mutually exclusive and exhaustive events is called a partition of the sample space. For instance, events A and A^c from a partition as they are mutually exclusive and their union is the entire sample space.

Equally Likely Events

Two events A and B Are said to be equally likely when one event is as likely to occur as the other. In other words, each event should occur in equal number in repeated trials. *For example*, when a fair coin is tossed, the head is as likely to appear as the tail, and the proportion of times each side is expected to appear is 1/2.

Similarly, if a die which is a perfect unloaded cube is rolled, then the face containing 2 dots is as likely to be on top as the face containing 5 dots. The same will be the case with any other pair of faces. In general if two events A and B occur in an experiment, then A and B are said to be equally likely events if each one of them has equal number of chances of occurrence.

The following definition of probability was given by a French Mathematician P.S. Laplace (1749-1827) and it has been accepted as a standard definition by mathematicians all over the world.

Events and Symbolic Representations

For convenience, the verbal statement of some events and their corresponding symbolic representations in sets are listed below:

Verbal Statement	Set Notation
Event A	A
Event A is impossible	A =
Event A is sure (certain)	A = S
Not-A (event A does not occur)	= S - A
Event A or event	A = S
Events A or event B	A
Events A and event B	A
Events A and B are mutually exclusive	A
Events A and B are exhaustive	A
Event B occurs when A occur	A
Event A occurs but B does not occur	A
Event B occurs given that A has occurred	B

Counting Sample Points

When the number of sample points in a sample space S is very large, it becomes very inconvenient and difficult to list them all and to count the number of points in the sample space S and in the subset of S. We then need some methods or rules which will help us to count the number of all sample points without actually listing them. A few of the basic rules frequently used in counting are briefly described here.

Rule of Multiplication

If a compound experiment consist of two experiments such that the first experiment has exactly m distinct outcomes and, if corresponding to each outcomes of the first experiment there can be n distinct outcomes of the second experiment, then the compound experiment has exactly *mn* outcomes.

For example, the compound experiment of tossing a coin and throwing a die together consists of two experiment: the coin-tossing with two distinct outcomes *(H, T)* and the die-throwing with six distinct outcomes (1, 2, 3, 4, 5, 6). The total number of possible distinct outcomes of the compound experiment is therefore 2 x 6 = 12, as each of the two outcomes of the coin-tossing can occur with each of the six outcomes of die-throwing experiment A tree diagram provides a good illustration of this rule.

The rule of multiplication can be readily extended to compound experiments consisting of any number of experiment performed in a given sequence.

Permutation

Suppose we like to find the number of different ways to name the triangle with vertices The various possible ways are obtained by constructing tree diagram

		B -------- C
	A	
		C---------B
Start	B	A---------C
		C---------A
	C	A--------B
		B---------A

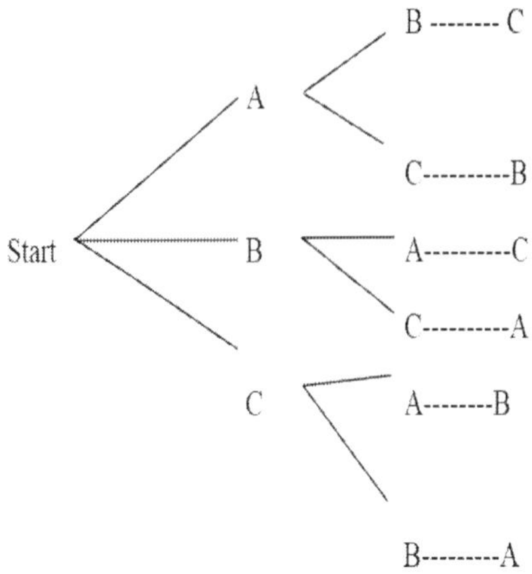

To determine the possible ways, we count the paths of the tree, beginning from the start to the end of each branch.

So we get 6 different names of triangle.

$ABC, ACB, BCA, BAC, CAB, CBA.$

Thus, there are six possible ways to write the name of the triangle with vertices

Explanation: In the above figure, we can write anyone of the three vertices at first place. After writing at first place anyone of the three vertices, two vertices are left. So, there are two choices to write at second place. After writing the vertices at two places, there is just one vertex left. So, we can write only one vertex at third place.

Another way of explanation: Think of the three places as shown ___, ___, ___. Since we can write any one of the three vertices at first place, so it is written in 3 different ways as shown **3, ___, ___**

Now two vertices are left. So, corresponding to each way of wining at first place, there are two ways of writing at second place as shown **3, 2, __.**

Now just one vertex is left. So we can write at third place only one vertex in one way as shown **3, 2, 1.**

The total number of possible ways (arrangements) is the product $3\times2\times1=6$.This example illustrates the fundamental principle of counting.

Fundamental Principle of Counting

Suppose $A\,and\,B$ are two events. The first event A can occur in p different ways. After A has occurred, B can occur in q different ways. The number of ways that the two events can occur is the product $p.q.$

This principle can be extended to three or more events. For instance, the number of ways that three events $A, B\,and\,C$ can occur in the product $p.q.r$.

One important application of the fundamental principle of counting is to determine the number of ways that n objects can be arranged in order. An ordering (arrangement) of n objects is called a permutation of the objects.

A permutation of n different objects is an ordering (arrangement) of the objects such that one object is first, one is second, one is third and so on.

According to the Fundamental Principle of Counting:

Three books can be arranged in a row taken all at a time $=3.2.1=3!$ ways

Number of ways of writing the letters of the word taken all at a time $=4.3.2.1=4!$

Each arrangement is called a permutation. Now we have the following definition.

A permutation of n different objects taken $r(\leq n)$ at a time is an arrangement of the r objects. Generally, it is denoted by

n_{Pr} or $P(n,r)$.

Rule of Permutation

A permutation is any ordered subset from a set of n distinct objects. The number of permutations of r objects, selected in a definite order from n distinct objects is denoted by the symbol n_{pr} $(r<n)$, , we proceed as below:

The first object may be chosen in n ways, and corresponding to each way of first selection, the second object may be chosen in (n - 1) (n -2)... (n-r+1) ways.

Hence, $np = n(n-1)(n-2)$ .... (n-r+1)

In particular, when $r = n, np = n(n-1)(n-2)\ ...3x2x1$

$= n!(read\ n\ factorial)$

It is relevant to note that 1! = 1 and that we defined 0! =1

The expression for $P_r\ on$ multiplication by $\frac{(n-r)!}{(n-r)!}$, may be written as $n_{pr} = \frac{n!}{(n-r)!}$

The number of permutations of n objects selected all at a time. when n objects consist of n_1 of one kind, n_2 of a second kind, ..., n_k of a kth kind, ($\sum n_i = n) is\ P = \frac{n!}{n_1!n_2 ....n_k^!}$.

Prove that $n_{P_r} = n(n-1)(n-2)...(n-r+1) = \frac{n!}{(n-r)!}$

Proof: As there are n different objects to fill up r places. So, the first place can be filled in n ways. Since repetitions are not allowed, the second place can be filled in $(n-1)$ ways, the third place is filled in $(n-2)$ ways and so on. The rth place has $n-(r-1) = n-r+1$ choices to be filled in. Therefore, by the fundamental principle of counting, r places can be filled by n different objects in $n(n-1)(n-2)...(n-r+1)$ ways.

$\therefore n_{P_r} = n(n-1)(n-2)...(n-r+1)$

$$= \frac{n(n-1)(n-2)...(n-r+1)(n-r)(n-r-1)...3.2.1}{(n-r)(n-r-1)...3.2.1}$$

$$n_{P_r} = \frac{n!}{(n-r)!}$$

which completes the proof.

Corollary: if $r = n,$ then

$$n_{P_n} = \frac{n!}{(n-n)!} = \frac{n!}{0!} = \frac{n!}{1} = n!$$

N different objects can be arranged taken all at a time in $n!$ ways.

***Example*:** How many different 4-digits numbers can be formed out of the digits $1,2,3,4,5,6$ when no digit is repeated?

***Solution*:** The total number of digits = 6

The digits forming each number = 4

So, the required number of 4- digit number is given by:

$$6_{P4} = \frac{6!}{(6-4)!} = \frac{6!}{2!} = \frac{6\times5\times4\times3\times2\times1}{2\times1} = 360$$

***Example*:** How many signals can be made with 4 different flags when any number of them are to be used at a time?

***Solution*:** The number of flags = 4

Number of signals using 1 flag = $4_{P1} = 4$

Number of signals using 2 flags = $4_{P2} = 12$

Number of signals using 3 flags = $4_{P3} = 24$

Number of signals using 4 flags = $4_{P4} = 24$

Total number of signals $= 4+12+24+24 = 64.$

***Example*:** In how many ways can a set of 4 different mathematics books and 5 different physics books be placed on a shelf with a space for 9 books, if all books on the same subject are kept together?

***Solution*:** 4 different Mathematics books can be arranged among themselves in $4!$ ways. 5 different Physics books can be arranged among themselves in $5!$ ways of arranging 5 physics books. The books in the two subjects can be arranged subject wise in 2! ways.

So the number of ways of arranging the books is given by.

$4!\times5!\times2!=4\times3\times2\times1\times5\times4\times3\times2\times1\times2\times1=5760$

Permutation of Things Not All Different

Suppose we have to find the permutation of the letters of the word BITTER using all the letters in it. We see that all the letters of the word BITTER are not different and it has 2 Ts in it. Obviously, the interchanging of Ts in any permutation, say BITTER, will not form a new permutation. However, if the two Ts are replaced by $T_1 \, and \, T_2$, we get the following two permutation of BITTER.

$BIT_1T_2ER \, and \, BIT_2T_1ER$

Similarly, the replacement of the two Ts by $T_1 \, and \, T_2$ in any other permutation will give rise to 2 permutations.

Now, BIT_1T_2ER consists of 6 different letters which can be permuted among themselves in 6! different ways. Hence the number of permutation of the letters of the word $BITTER$ taken all at a time

$$=\frac{6!}{2}=\frac{6\times5\times4\times3\times2\times1}{2}=360$$

This example guides us to discover the method of finding the permutation of n things all of which are not different. Suppose that out of n things, n_1 are alike of one kind and n_2 are alike of second kind and the rest of them are all different. Let x be the required number of permutations. Replacing n_1 alike things by n_1 different things and n_2 alike things by n_2 different things, we shall get all the n things distinct from each other which can be permuted among themselves in $n!$ ways. As n_1 different things can be permuted among themselves in $(n_1)!$ ways and n_2 different things can be arranged among themselves in $(n_2)!$ ways. So because of the replacement suggested above, x permutation would increase to

$x\times(n_1)!\times(n_2)!$ number of ways.

$\therefore x \times (n_1)! \times (n_2)! = (n)!$

Hence $$x = \frac{(n)!}{(n_1)! \times (n_2)!} = \binom{n}{n_1, n_2}$$

Corollary: If there are n_1 alike things of one kind, n_2 alike things of second kind and n_3 alike things of third kind, then the number of permutation of n things taken all at a time is given by:

$$\frac{n!}{(n_1)! \times (n_2)! \times (n_3)!} = \binom{n}{n_1, n_2, n_3}$$

Example: In how many ways can be letters of the word MISSISSIPPI be arranged when all the letters are to be used?

Solution: Number of letters in $MISSISSIPPI = 11$

In MISSISSIPPI,

I is repeated 4 times

S is repeated 4 times

P is repeated 2 times

M comes only once.

Required number of permutation

$$= \binom{11}{4,4,2,1}$$

$$= \frac{(11)!}{(4)! \times (4)! \times (2)! \times (1)!} = 34650\, ways$$

Circular Permutation

So far we have been studying permutation of things which can represented by the points on a straight line. We shall now study the permutation of things which can be represented by the points on a circle. The permutation of things which can be represented by the points on a circle are called Circular Permutation.

The method of finding circular permutation is illustrated by the following examples.

***Example*:** In how many ways can 5 persons be seated at a round table?

***Solution*:** Let A, B, C, D, E be the 5 persons. One of the ways of seating them round a table is shown in the adjoining figure.

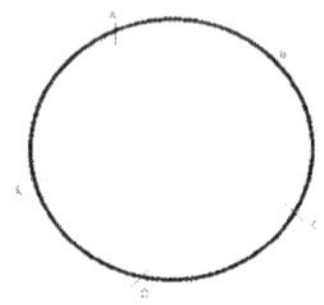

If each person moves one or two or more places to the left or the right, they will no doubt, be occupying different seats, but their positions relative to each other will remain the same.

So, when A occupies a certain seat, the remaining 4 persons will be permuting their seats among themselves in 4! ways.

Hence the number of arrangements = 4! = 24

***Example*:** In how many ways can a necklace of 8 beads of different colors be made?

***Solution*:** The number of beads = 8

The number of arrangements of 8 beads in the necklace will be like the seating of 8 persons round a table. The number of such necklace (fixing one of the beads) = 7!

Now suppose the beads are a, b, c, d, e, f, g, h and the necklace is as shown in figure.

By flipping the necklace we get the necklace as shown in figure.

We observe that the two arrangements of the beads are actually the same

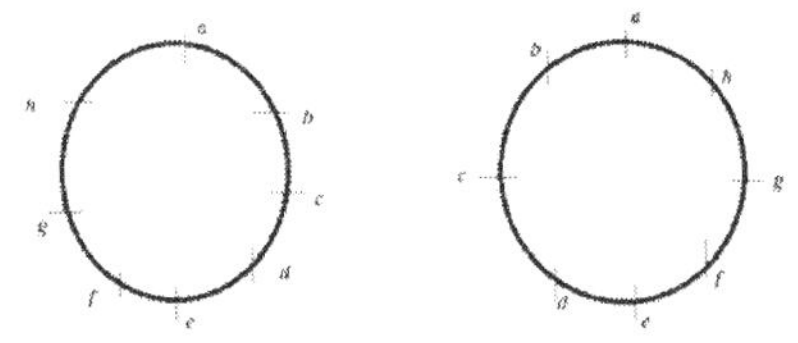

Hence the required number of necklaces $= \frac{1}{2} \times (7)! = 2520$

Combinations

While counting the number of possible permutation of a set of objects, the order is important. But there are situations where order is immaterial. For example,

$ABC, ACB, BAC, BCA, CAB, CBA$ are the six names of the triangle whose vertices are $A, B \, and \, C.$ We notice that in spite of the different arrangements of the vertices of the triangle, they represent one and the same triangle.

The 11 players of a cricket team can be arranged in 11! ways but they are players of the same single team.

So, we are interested in the membership of the committee (group) and not in the way the members are listed (arranged). Therefore, a combination of n different objects taken r at a time is a set of r objects.

The number of combination of n different objects taken r at a time is denoted by $n_{C_r} \, or \, C(n,r) \, or \binom{n}{r}$ and is given by $n_{C_r} = \frac{n!}{r!(n-r)!}$

Rule of Combination

A combination is any subset of r objects, selected without regard to their order, from a set of n distinct objects. The total number of such combinations is denoted by the symbol n_{C_r} or $\left(\frac{n}{r}\right)$, read n above r, where r $\leq n.$

The computational formula for Combination is derived as below:

Let there be $\left(\frac{n}{r}\right)$ *possible combinations* of r objects, selected from n distinct objects. Then each combinations of r objects, if order is not disregarded, can be arranged in $r^!$ different orders and thereby makes $r^!$ different permutations. The number of permutations or orders subsets that correspond to $\left(\frac{n}{r}\right)$ combinations, each having r objects, selected from n objects is nPr. Thus, we conclude that

$$r!\left(\frac{n}{r}\right) = n_{Pr}$$

or

$$\left(\frac{n}{r}\right) = \frac{\text{n}P_r}{r\,!} = \frac{n(n-1)(n-2)\ldots.(n-r+1)}{r(r-1)(r-2)\ldots321}$$

$$= \frac{n\,!}{r\,!(n-r)!}$$

The quality $\left(\frac{n}{r}\right)$ or n_{Cr} is also called a binomial coefficient because of its appearance in the binomial expansion of $(a+b)^n = \sum_{r=0}^{n}\left(\frac{n}{r}\right)a^{n-r}b^r$

The binomial coefficient has two important properties.

(i) $\left(\frac{n}{r}\right) = \left(\frac{n}{n-r}\right)$

(ii) $\left(\frac{n}{r-1}\right) + \left(\frac{n}{r}\right) = \left(\frac{n+1}{r}\right)$

Large factorials may be conveniently evaluated by using an approximation known as Stirling's formula:

$$n! \cong n^n + 1/2e^{-n}\sqrt{2\pi},$$

where $\pi = 3.1416$ and $e = 2.7183$

Complementary Combination

Prove that $n_{Cr} = n_{cn-r}$

Proof: If from n different objects, we select r objects then $(n-r)$ objects are left. Corresponding to every combination of r objects, there is a combination of $(n-r)$ objects and vice versa.

Thus the number of combinations of n objects taken r at a time is equal to the number of combination of n objects taken $(n-r)$ at a time.

$$\therefore n_{cr} = n_{cn-r}$$

Otherwise, $n_{cn-r} = \dfrac{n!}{(n-r)!(n-r+r)!} = \dfrac{n!}{(n-r)!r!} = \dfrac{n!}{r!(n-r)!}$

$$\Rightarrow n_{cn-r} = n_{cr}$$

Note: This result will be found useful in evaluating n_{cr} $when\, r > \dfrac{n}{2}$.

e.g. $12_{c10} = 12_{c12-10} = 12_{c2} = \dfrac{(12)\times(11)}{2} = 6\times 11 = 66$

Example: If $n_{C8} = n_{C12}$, find n.

Solution: We know that $n_{Cr} = n_{Cn-r}$

$n_{C8} = n_{Cn-8}$ (i)

But it is given that $n_{C8} = n_{C_{12}}$ (ii)

So from (i) and (ii), we conclude that

$n_{Cn-8} = n_{C12}$

$\Rightarrow n-8 = 12$

$\therefore n = 20$

Example: Find the number of the diagonal of a 6 sided figure.

Solution: A 6-sided figure has 6 vertices. Joining any two vertices we get a line segment

$$\therefore Number\ of\ line\ segments = 6_{C2} = \frac{6!}{2!4!} = 15$$

But these line segments include 6 sides of the figure

$$\therefore Number\ of\ diagonlas = 15 - 6 = 9$$

***Example*:** A club consists of four members. How many sample points are in the sample space when three officers: president, secretary and treasurer, are to be chosen?

***Solution*:** It is evident that the order in which 3 officers are to be chosen, is of significance. Thus there are 4 choices for the third office. Hence the number of sample points is 4 x 3 x 2 = 24. In other words, the number of permutations is

$$_4P_3 = \frac{4!}{4-3\,!} = 4 \text{ x } 3 \text{ x } 2 = 24$$

Let the four members be A, B, C and D. Then a tree diagram that provides an organized way of listing the possible arrangements, for this example, is given below:

President Secretary Treasurer Sample Space

ABC
ABD
ACB
ACD
ADB
ADC
BAC
BAD
BCA
BCD
BDA
BDC
CAB

CAD
CBA
CBD
CDA
CDB
DAB
DAC
DBA
DBC
DCA
DCB

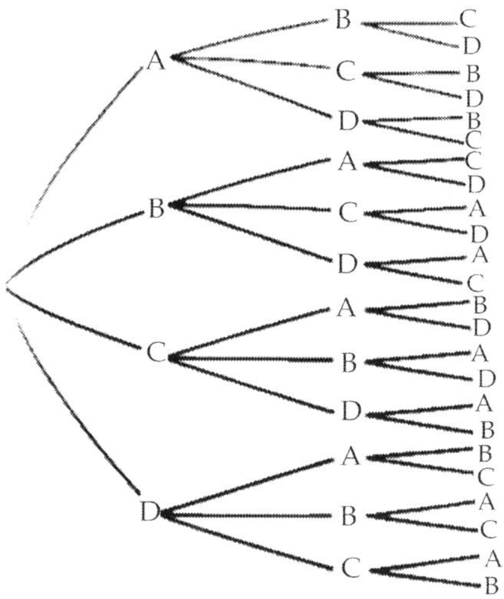

***Example*:** A three-person committee is to be formed from a list of four persons. How many sample points are associated with the experiment?

***Solution*:** Since the order in which the three persons of the committee are chosen is unimportant, it is therefore an example of a problem involving combinations. Thus the desired number of combinations is

$$\binom{n}{r} = \binom{4}{3} = \frac{4\,!}{3\,!\,(4-3)!} = 4$$

In other words, the sample space associated with the experiment contains four sample points.

These two examples serve to illustrate the difference between a permutation and a combination.

***Example*:** How many sample points are in the sample space when a person draws a hand of five cards from a well-shuffled ordinary deck of 52 cards?

***Solution*:** The total number of sample points is given by

$$\left(\frac{n}{r}\right) = \left(\frac{52}{5}\right) = \frac{52\times51\times50\times49\times48}{5\times4\times3\times2\times1} = 2,598,960$$

Probability

We live in an uncertain world where very many events cannot be predicted with complete certainty. For example,

In a cloudy weather, we cannot be sure whether it will or will not rain. However, we can say that there is 1 chance out of 2 that the rain will fall.

There are 6 theorems on circle out of which one theorem is asked in the secondary school examination. Evidently, there is 1 chance out of 6 that a particular theorem will be asked in the examination.

In simple situations, we are guided by our experience or intuition. However, we cannot be sure about our predictions. Nevertheless, in more complex situations, we cannot depend upon guess work and we need more powerful tools for analyzing the situations and adopting the safer path for the achievement of our goals. In order to guide in solving complex problems of everyday life, two French mathematician, Blaise Pascal (1623- 1662) and Pierre de Fermat (1601- 1665), introduced probability theory. A very simple definition of probability is given below:

Probability is the numerical evaluation of a chance that a particular event would occur.

This definition is too vague to be of any practical use in estimating the chance of the occurrence of a particular event in a given situation.

Definitions of Probability

Probability can be discussed from two points of view: the objective and the subjective. The objective probability can be classified into the following categories, each of which is briefly described as follows:

(a) ***The Classical or A Priori Definition of Probability*** is given as follows:

If a random experiment can be produced in n mutually exclusive and equally likely outcomes, and if m out of these outcomes are considered favorable to the occurrence of a certain event A, then the probability of the event A, denoted by P(A), is defined as the ratio $\frac{m}{n}$. Symbolically, we write

$$P(A)=\frac{m}{n}=\frac{Number\ of\ favorable\ outcomes}{Total\ number\ of\ possible\ outcomes}$$

This definition was formulated by the French mathematician P.S. Laplace (1749-1827) and can be very conveniently used in experiment where the total number of possible outcomes and the number of outcomes favorable to an event can be determined.

The classical definition has the following shortcomings:

(i) This definition is said to involve circular reasoning as the term equally likely really means equally probable. Thus, probability is defined by introducing concepts that presume a prior knowledge of the meaning of probability.

(ii) This definition is not applicable when the assumption of equally likely does not hold.

(iii) This definition becomes vague when the number of possible outcomes may be infinite.

(b) *The Relative Frequency or A Posteriori Definition or Probability*. If a random experiment is repeated a large number of times say n, under identical conditions and if an event A is observed to occur m times, then the probability of the event A is defined as the limit of the relative frequency $\frac{m}{n}$ as n tends to approach infinity. Symbolically, we write

$$`P(A)=lim_{n \to \infty} \frac{m}{n}$$

This definition assumes that as n increases indefinitely, the ratio $\frac{m}{n}$ tends to become stable at the numerical value P(A). To investigate this kind of long-run stability, several coin-tossing experiments have been performed. The result of a few well-known experiments is shown below:

Experimenter	Number of times coin tossed (n)	Number of heads obtained (m)	Ratio $\frac{m}{n}$
Buffon	4,040	2,048	0.5069
K. Pearson	12,000	6,019	0.5016
K. Pearson	24,000	12,012	0.5005

It is obvious that the value of the ratio $\frac{m}{n}$ fluctuates about the number 0.5 and become closer to 0.5 as the number of throws increases. This sort of a long-run frequency property provides as useful basis of the theory of probability.

This definition is also called the statistical or empirical definition of probability as it is based on observational data. It is more useful for practical problems.

This definition too has certain limitations as the conditions under which an experiment is performed, may change from trial to trial and it is not possible, in practice, to repeat the experiment an infinite number of times and hence the ratio $\frac{m}{n}$ may not be unique.

(c) **The Axiomatic Definition of Probability.** This definition, introduced in 1933 by the Russian mathematician Andrie N. Kolmogorov (1903--), is based on a set of axioms, where an axiom is a statement that is assumed to be true. Let S be a sample Space with the Sample points $E_1, E_2, \ldots E_i \ldots E_n$. We assign a real number to each sample point denoted by the symbol P(E_i),and called the probability of E_i, that must satisfy the following basic axioms:

Axiom (i) For any event $E_i, 0 \le P(E_i) \le 1$

Axiom (ii) P(S) = 1 for the sure event S

Axiom (iii) if A and B are mutually exclusive events (subsets), then P $(A \cup B)$ = P (A) + P(B).

It is to be emphasized that the axiomatic theory of probability assumes that some probability defined as a non-negative real number is to be attached to each sample point E_i such that the sum of all such number must be equal to one. The assignment of probabilities may be based on past evidence or on some other underlying conditions.

Probability of An Event A

If an event A is defined in a sample space S, then its probability will be equal to the sum of the probabilities of all sample points that are included in A, i.e. P(A) = $\Sigma P(E_i)$. When all the n possible outcomes is assigned the same probabilities $\frac{1}{n}$, e.g. in throwing a fair die once,

P(1) = $\frac{1}{6}, P(2) = \frac{1}{6}, \ldots P(6) = \frac{1}{6}$

It follows from axiom (iii) that for any event A containing m equally likely outcomes (sample points), we have

(i) $P(A) = \frac{m}{n} \ge 0;$

(ii) $P(S) = \frac{m}{n} = 1;$

(iii) If A and B are mutually exclusive events, and m_1 *and* m_2 outcomes are favorable to A and B, then $P(A \cup B) = \frac{m_1 + m_2}{n} = P(A) + P(B)$

Thus, the classical definition of probability also satisfies Kolomogorv's axioms. It is reasonable to conclude that the probability is always a number between zero and one (inclusive).

Subjective or Personalistic Probability

As its name suggest, the subjective or personalistic probability is a measure of the strength of a person's belief regarding the occurrence of an event A. Probability in this sense is purely subjective and is based on whatever evidence is available to the individual. This definition being flexible, may be applied to those real-world situations where neither an equally likely nor a relative frequency approach is possible. The subjective probability has a disadvantage that two or more person faced with the same evidence may arrive at different probabilities.

Example: If a card is drawn from an ordinary deck of 52 playing cards, find the probability that (i) the card is a red card, (ii) the card is diamond, and (iii) the card is a 10.

Solution: The total number of possible outcomes is 52, and we assume that all possible outcomes are equally likely.

(i) Let A represent the event that the card drawn is a red card. Then the number of outcomes favorable to the event A is 26 since there are 26 red cards

$$Hence\, P(A)=\frac{m}{n}=\frac{Number\ of\ favorable\, outcomes}{Total\ Numbers\ of\ possible\, outcomes}$$

$$=\frac{26}{52}=\frac{1}{2}$$

(ii) Let B denote the event that the card drawn is a diamond. Then the number of outcomes favorable to the event B is 13 since there are 13 diamonds.

$$Hence\, P(B)=\frac{13}{52}=\frac{1}{4}$$

(iii) Let C denote the event that the card drawn is a 10. Then the number of outcomes favorable to C is 4 as there are four 10's.

$$Thus, P(C)=\frac{4}{52}=\frac{1}{13}$$

Example: A fair coin is tossed three times. What is the probability that at least one head will appear?

Solution: The sample space for this experiment is

$S=\{HHH, HHT, HTH, THH, HTT, THT, TTH, TTT\}$

Thus $n(S)=8$. As the coin is fair, we assume that each of these are equally likely to appear. Therefore, we assign equal probability of $\frac{1}{8}$ to each outcome.

Let A denote the event that at least one head appear. Then

$A=\{HHH, HHT, HTH, THH, HTT, THT, TTH\}$

Therefore, $n(A)=7.$

Hence
$$P(A)=\frac{n(A)}{n(S)}=\frac{7}{8}$$

Example: If two fair dice are thrown, what is the probability of setting (i) a double six? (ii) a sum of 8 or more dots?

Solution: The sample space S is represented by the following 36 outcomes:

S = { (1, 1), (1, 2), (1, 3), (1, 4), (1, 5), (1, 6)
(2, 1), (2, 2), (2, 3), (2, 4), (2, 5), (2, 6)
(3, 1), (3, 2), (3, 3), (3, 4), (3, 5), (3, 6)
(4, 1), (4, 2), (4, 3), (4, 4), (4, 5), (4, 6)
(5, 1), (5, 2), (5, 3), (5, 4), (5, 5), (5, 6)
(6, 1), (6, 2), (6, 3), (6, 4), (6, 5), (6, 6)}

As the dice are fair, therefore each of these 36 outcomes are equally likely and a probability of $\frac{1}{36}$ is attached with each outcomes.

(i) Let A represent the event that a double six occurs.

Then A = {6, 6} and thus P(A) = $\frac{1}{36}$

(iii) Let B denote the event that a sum of 8 or more dots occurs

Then B = { (6, 2), (5, 3), (4, 4), (3, 5), (2, 6), (6, 3), (5, 4), (4, 5), (3,6), (6,4) (5,5), (4,6), (6,5), (5, 6), (6, 6) }

Hence P(B) = $\frac{15}{36} = \frac{5}{12}$

Example: Six white balls and four black balls, which are indistinguishable apart from color, are placed in a bag. If six balls are taken from the bag, find the probability of there being three white and three black.

Solution: The total number of possible equally likely outcomes in S is

$$\binom{10}{6} = \frac{10\,!}{6\,!(10-6)!} = 210$$

Let A represent the event that three white and three black balls are taken.

Then the number of outcomes that correspond to the event A is $\binom{6}{3}$ x $\binom{4}{3} = 80$

Therefore P(A) = $\frac{n(A)}{n(S)} = \frac{80}{210} = \frac{8}{21}$,

Example: An employer wishes to hire three people from a group of 15 applications, 8 men and 7 women are equally qualified to fill the position. If he selects three at random, what is the probability that (i) all three will be men, (ii) at least one will be a woman.

Solution: The sample space S contains $\binom{15}{3}$ = 455 sample points the number of way in which 3 people can be selected from 15 applicants.

(i) Let A represent the event that the three selected will be men. Then A contains $\binom{8}{3} = 56$ sample points, the number of ways in which 3 men can be selected from 8 men.

$$\text{Therefore}, P(A) = \frac{n(A)}{N(S)} = \frac{56}{455} = \frac{8}{65}$$

(ii) "At least one woman" means one, two or three women. Let B denote the event that at least one woman is selected.

$$\text{Then}, n(B) = \binom{7}{1}\binom{8}{2} + \binom{7}{2}\binom{8}{1}\binom{7}{3}$$

$$= 196 + 168 + 35 = 399 \text{ sample points}$$

$$\text{Hence}, P(B) = \frac{n(B)}{N(S)} = \frac{399}{455} = \frac{57}{65} = 0.877$$

Example: Four items are taken at random from a box of 12 items and inspected. The box is rejected if more than 1 items is found to be faulty. If there are 3 faulty items in the box, find the probability that the box is accepted.

Solution: The sample space S contains $\binom{12}{4} = 495$ sample points. The box contains 3 faulty and 9 good items. The box is accepted if there is (i) no faulty items, or (ii) one faulty items in the sample of 4 items selected.

Let A denote the event that the number of faulty items chosen is 0 or 1. Then,

$$n(A) = \binom{3}{0}\binom{9}{4} + \binom{3}{1}\binom{9}{3}$$

$$= 126 + 252 = 378 \text{ sample points}$$

$$P(A) = 378/495 = 0.76$$

Hence the probability that the box is accepted is 0.76.

SUMMARY AND DISCUSSION

In this chapter, we discussed some elementary aspects of probability theory. A probability problem can usually be broken down into a few basic steps:

The description of the sample space, that is, the set of possible outcomes of a given experiment.

The arrangement of the probability law (the probability of each event). The computation of probabilities and conditional probabilities of various events of importance. The probabilities of events must assure the no negativity, additive, and normalization axioms. In the significant special case where the set of possible outcomes is predetermined, one can just state the probability of each outcome and obtain the probability of any event by adding the probabilities of the elements of the event.

Given a probability law, we are often concerned with conditional probabilities, which permit us to reason based on partial information about the outcome of the conducted experiment. We can view conditional probabilities as probability laws of a special type, under which only outcomes restricted in the conditioning event can have positive conditional probability. Conditional probabilities can be derived from the probability law using the definition $P(A|B)=P(A n B)/P(B)$. However, the overturn procedure is frequently suitable, that is, first identify some conditional probabilities that are likely for the real circumstances that we wish to model, and then apply them to obtain the probability law. Through examples, we have illustrated three methods for calculating probabilities:

(a) *The Counting Method.* This method applies to the case where the number of probable outcomes is finite, and all outcomes are equally likely. To compute the probability of an event, we count the number of elements of the event and divide by the number of elements of the sample space.

(b) *The Sequential Method.* This method applies when the experiment has a chronological character, and appropriate conditional probabilities are specified or calculated along the branches of the corresponding tree. The probabilities of different events are then obtained by multiplying conditional

probabilities along the corresponding paths of the tree, using the multiplication rule.

Sets

A set is any well-defined collection or list of distinct objects, e.g. a group of students, the books in a library, the integers between 1 and 100 all human beings on the Earth, etc.

Relation and Function

A relation from a set A to set a subset of the Cartesian product of A x B. Such a relation is usually called a binary relation. That is a relation is an association between two or more objects. The set of the first elements of a binary relation is called the domain of the relation while the set of the second elements of the relation is called the range. For example, if a relation in relation is called the range. For example, if a relation F = {(1, 4), (2, 7), (3, 12)}. Then its domain is D = {1, 2, 3} and its range R = {4, 7, 12}.

Random Experiment

The term experiment means a planned activity or process whose result yield a set of data. A single performance of an experiment or a trial is called a trial. The result obtained from an experiment or a trial is called an outcome.

An experiment which produces different results even though it is repeated a large number of times under essentially similar conditions, is balanced die. Drawing of a card from a well-shuffled deck of 52 playing cards. A random experiment has three properties:

(i) the experiment can be repeated, practically or theoretically, any number of times;

(ii) the experiment always has two or more possible outcomes. An experiment that has only one possible outcomes, is not a random experiment;

(iii) the outcome of each repetition is unpredictable, i.e. it has some degree of uncertainty.

Sample Space

A set consisting of all possible outcomes that can result from a random experiment (real or conceptual), is defined to be a sample for the experiment (and is denoted by the letter S). Each possible outcome is a member of the sample space and is called a sample point in that space.

Events

An event is an individual outcome or any number of outcomes (sample points) of a random experiment or a trial. In set terminology, any subset of a sample space S of the experiment is called an event. An event that contains exactly one sample point is defined as a simple event. A compound event contains more than one sample point and is produced by the union of sample events. For instance, the occurrence of a sum of 10 with a pair of dice, is a compound event, as it can be decomposed into three simple events (4, 6), (5, 5) and (6, 4)

An event A is said to occur if and only if the outcome of the experiment corresponds to some element of A. The event " not-A" is denoted by A or A^c and is called the negation (or complementary event) of A.

Mutually Exclusive Events

Two events A and B of a single experiment are said to be mutually exclusive or disjoint and only if they cannot both occur at the same time. That is they have no points in common For instance, when we toss a coin, we get either a head or a tail, but not both, the two events head and tail are therefore mutually exclusive; when a die is rolled, the outcomes are mutually exclusive as we get one and only one of six possible outcomes 1,2,3,4,5 or 6.

Exhaustive Events

Events are said to be collectively exhaustive, when the union of mutually exclusive events is the entire sample space S. Thus, in our coin-tossing experiment, head and tail are collectively exhaustive set of events. A group of mutually exclusive and exhaustive events is called a partition of the sample space. For instance, events A and A^c from a partition as they are mutually exclusive and their union is the entire sample space.

Equally Likely Events

Two Events A and B are said to be equally likely, when one event is as likely to occur as the other. In other words, each event should occur in equal number in repeated trials. For example, when a fair coin is tossed, the head is as likely to appear as the tail, and the proportion of times each side is expected to appear is 1/2.

Rule of Multiplication

If a compound experiment consists of two experiments such that the first experiment has exactly m distinct outcomes and, if corresponding to each outcomes of the first experiment there can be n distinct outcomes of the second experiment, then the compound experiment has exactly *mn* outcomes.

Rule of permutation

A permutation is any ordered subset from a set of n distinct objects. The number of permutations of r objects, selected in a definite order from n distinct objects is denoted by the $ymbol\ \ n_{pr}$.

Rule of Combination

A combination is any subset of r objects, selected without regard to their order, from a set of n distinct objects. The total number of such combinations is denoted by the symbol C_r or $\left(\frac{n}{r}\right)$, read n above r, where r $\leq n$.

The Classical Definition of Probability

If a random experiment can be produced in mutually exclusive and equally likely outcomes and if m out of these outcomes are considered favorable to the occurrence of a certain event A, then the probability of the event A, denoted by P(A), is defined as the ratio $\frac{m}{n}$. Symbolically, we write

$$P(A)=\frac{m}{n}=\frac{Number\ of\ favorable\ outcomes}{Total\ number\ of\ possible\ outcomes}$$

The classical definition has the following short comings:

(i) This definition is said to involve circular reasoning as the term equally likely really means equally probable. Thus probability is defined by introducing concepts that presume a prior knowledge of the meaning of probability.

(ii) This definition is not applicable when the assumption of equally likely does not hold.

(iii) This definition becomes vague when the number of possible outcomes may be infinite.

The Relative Frequency or *A Posteriori* Definition or Probability

If a random experiment is repeated a large number of times say n, under identical conditions and if an event A is observed to occur m times, then the probability of the event A is defined as the limit of the relative frequency $\frac{m}{n}$ as n tends to infinity. Symbolically, we write

$$`P(A)=lim_{n\to\infty}\frac{m}{n}$$

This definition assumes that as n increases indefinitely, the ratio $\frac{m}{n}$ tends to become stable at the numerical value P(A).

The Axiomatic Definition of Probability

Let S be a sample space with the sample points $E_1, E_2, \ldots E_i \ldots E_n$. To each sample point, we assign a real number, denoted by the symbol P(E_i),and called the probability of E_i, that must satisfy the following basic axioms:

Axiom (i) for any event E_i, $0 \leq P(E_i) \leq 1$

Axiom (ii) P(S) = 1 for the sure event S

Axiom (iii) if A and B are mutually exclusive events (subsets), then P (A $\cup B$) = P (A) + P(B).

Probability of an event A

If an event A is defined in a sample space S, then its probability will be equal to the sum of the probabilities of all sample point that are included in A, i.e. P (A) = $\Sigma P(E_i)$.When all the n possible outcomes is assigned the same probabilities $\frac{1}{n}$,, e.g. in throwing a fair die once, P(1) = $\frac{1}{6}, P(2)=\frac{1}{6},\ldots P(6)=\frac{1}{6}$,it follows from axiom (iii) that for any event A containing m equally likely outcomes (sample points) we have

(i) $P(A)=\frac{m}{n}\geq 0;$

(ii) $P(S)=\frac{m}{n}=1;$

(iii) If A and B are mutually exclusive events, and m_1 *and* m_2 *outcomes* are favorable to A and B, then $P(A\cup B)=\frac{m_1+m_2}{n}=P(A)+P(B)$.

CHAPTER 2

AXIOMS OF PROBABILITY

INTRODUCTION

In the previous chapter we discussed about the basic definitions and rules of probability. In this chapter we will discuss axioms of Probability in detail.

LAWS OF PROBABILITY

Following are some of the basic rules for the calculations of probability. These rules have important applications.

Theorem: If $\varnothing$ is the impossible event, then P($\varnothing$) = 0.

***Proof*:** The sure event S and the impossible events $\varnothing$ are mutually exclusive and their union is

$$S \cup \varnothing = S$$

$$Then\, P(S) = P(S \cup \varnothing) = P(S) + P(\varnothing)$$

Subtracting P(S) from both sides, we get

$$P(\varnothing) = 0$$

Thus, the probability of the impossible event is zero. It is to be kept in mind that the converse of this rule is not generally true. Moreover, a theorem is a statement derived either from axioms or previously proved theorems.

Law of Complementation

If $\bar{A}$ is the complement of an event A relative to the sample space S, then $P(\bar{A}) = 1 - P(A)$.

Proof. Since the event A and A are exclusive and collectively exhaustive, i.e. they together make up the entire sample space S, therefore, we have

A$\cup A = S$

Thus P(A $\cup A) = P(S)$

Or P(A) + P(A) =1 [P(S) = 1 by axiom (ii)]

Hence the probability of the complement of an event is equal to one minus the probability of the event. Complementary probabilities are very useful when the question asks for the probability of at least one event.

Example: A coin is tossed 4 times in succession. What is the probability that at least one head occurs?

Solution: The sample Space S for this experiment consist of 2^4 =16 sample points, as each toss can result in 2 outcomes is equally likely,

Let A represent the event that at least one head occurs. Then A consists of many sample points. On the other hand, A is the event that no head occurs and A has the single sample point {TTTT}. Therefore P(A) = $\frac{1}{16}$.

Hence by the law of complementation, we have

P(A) = $1 - P(A)$ $= 1 - \frac{1}{16} = \frac{15}{16}$.

Example: A coin is biased so that the probability that it falls is $3/4$

(a) Find the probability of obtaining at least one head when the coins is tossed five times.

(b) How many times must the coin be tossed so that the probability of obtaining at least one head is greater than 0.98?

Solution: Here, P(a head appears) = $\frac{1}{4}$, and

P(no head or a tail appears) = $\frac{3}{4}$.

(a) Let A be the event that at least one head is obtained when the coin is tossed 5 times, and A is the event that no head is obtained. Then by the law of complementation, we have

$$P(A)=1-P(\overline{A})$$

$$=1-\left(\frac{3}{4}\right)^5=1-0.237=0.763$$

(b) Let the coin be tossed n times so that the probability of obtaining at least one head is greater than 0.98.

Then $1-\left(\frac{3}{4}\right)^n \geq 0.98 \; i.e \left(\frac{3}{4}\right)^n \geq 0.02$

Taking logs, we have

$$n.log\left(\frac{3}{4}\right)\geq \log 0.02$$

Dividing both sides by log $\left(\frac{3}{4}\right)$ and reversing the inequality sign as log $\left(\frac{3}{4}\right)$ is negative, we have

$$n \geq \frac{\log 0.0.2}{\log\left(\frac{3}{4}\right)} \geq \frac{\bar{2}.3010}{\bar{1}.8751} \geq \frac{-1.6990}{-0.1249} \geq 13.6$$

So at least n = 14.

Hence the coin should be tossed 14 times so that the probability of obtaining at least one head is greater than 0.98.

Probability of Sub Event

If A and B are two events such that A $\subset B$, then $P(A)\leq P(B)$.

Proof: For $A\subset B$, the event B may be written as the union of two mutually exclusive events

B $\cap A$ and B $\cap \bar{A}$,

i.e. $B=(B\cap A)\cup(B\cap \bar{A})$

But $B\cap A = A,\ so\ B = A\cup(B\cap \bar{A})$

Therefore $P(B)=P(A)+P(B\cap \bar{A})$

But $P(B\cap \bar{A})\geq 0.$

Hence P(A) $\leq P(B).$

It is to be noted that an event such as $A\cap B$ is called a joint event, and the probability associated with a joint equation is called a joint probability.

Theorem: If A and B are any two events defined in a sample space S. Then $P(A\cap\bar{B}) = P(A) - P(A\cap B)$

Proof: The events $A\cap\bar{B}$ and $A\cap B$ are mutually exclusive and their union is A. That is

$A=(A\cup\bar{B})\cup(A\cup B).$

$P(A)=P(A\cap\bar{B})+P(A\cap B)$

Here $P(A\cap\bar{B}) = P(A) - P(A\cap B).$

Addition Law

If A and B are any two events defined in a sample space S, then

$P(A\cup B)=P(A)+P(B)-P(A\cap B)$

Proof. The event A $\cup B$ may be written as the union of two mutually exclusive events A and $(B\cap\bar{A})$. That is $A\cup B = A\cup(B\cap \bar{A})$ then

$P(A\cup B)= P(A)+P(B\cap \bar{A})$

Again the event B may also be decomposed into two mutually exclusive events as

$$B=(A\cap B)\cup(\bar{A}\cap B)$$

Thus $P(B)=P(A\cap B)+P(\bar{A}\cap B)$

Subtracting this result from the former, we get

$$P(A\cup B)-P(B)=P(A)-P(A\cup B)$$

Hence $P(A\cup B)=P(A)+P(B)-P(A\cap B)$

This law, often called the **General Rule of Addition for Probabilities** may be stated as below:

"If two events A and B are not mutually exclusive, then the probabilities that at least one of them occurs, is given by the sum of the separate probabilities of events A and B minus the probability of the joint event $A\cap B$."

Corollary 1. If A and B are mutually exclusive events, then $P(A\cup B)=P(A)+P(B)$

Proof: Since the events A and B are mutually exclusive events, therefore A $\cap B$ = $\varnothing$ and

$$P(A\cap B)=P(\varnothing)=0$$

Hence $(A\cap B)=P(A)+P(B)$, which is just a restatement of axiom (iii).

Alternatively. Let n be the total number of sample points, and let A and B be two events; A consisting of m_1 sample points and B of m_2 sample points. The occurrence of A $\cup B$ is a set which consist of all the sample points belonging to either A or B

Since A and B are mutually exclusive events, therefore they have no sample points in common. Then, obviously, the number of points contained in A $\cup B\, is\, m_1+m_2$.

Hence P(A $\cup B) = \frac{number\ of\ sample\ points\ in\ A \cup B}{number\ of\ sample\ points\ in\ S}$

$= \frac{m_1 + m_2}{n} = \frac{m_1}{n} + \frac{m_2}{n}$ = P(A) + P(B).

Corollary 2. If $A_1, A_2, \ldots, A_k\ mu$ tually exclusive events, then the probability that one of them occurs, is the sum of the probabilities of the separate events, i.e.

P($A_1 \cup B_2 \cup \ldots\ldots \cup A_k) = P(A_1) + P(A_2) + \ldots\ldots, + P(A_k)$.

It is to be noted that if A_1, A_2, ... A_k are mutually exclusive and collectively exhaustive, the P($A_1) + P(A_2) + \ldots.. + P(A_k) = 1$.

Corollary 3. If A and B are any two events, then

P(A $\cup B) \leq P(A) + P(B)$

Proof: The addition law for any two events A and B may be written as

P(A $\cup B) + P(A \cap B) = P(A) + P(B)$

Therefore, $P(A \cup B) \leq P(A) + P(B)$

In general, for any k events $A_1, A_2, \ldots. A_k$, the relation

$P(A_1 \cup A_2 \cup \ldots \cup A_K) \geq P(A_1) + (A_2) + \ldots. + P(A_k)$

holds. This result is known as **Boole's inequality**.

***Example*:** If one card is selected is at random from a deck of 52 playing cards, what is the probability that the card is a club or a face card or both?

***Solution*:** Let A represent the event that the card selected is a club, and B is the event that the card selected is a face card. Then we need P(A $\cup B)$, the event that the card selected is both a club and a face card.

Now P(A) = $\frac{13}{52}$, as there are13 clubs

P(B) = $\frac{12}{52}$, as there are 12 face cards

and P(A $\cup B) = \frac{3}{52}$, since3 of clubs are also face cards

Therefore, the desired probability is

P(A $\cup B)$ = P(A) + P(B) – P(A $\cup B)$

$\frac{13}{52}+\frac{12}{52}-\frac{3}{52}=\frac{22}{52}$

Example: An integer is chosen at random from the first 200 positive integers. What is the probability that the integer chosen is divisible by 6 or by 8?

Solution: The sample space for this experiment is

S= { 1, 2, 3,……. 199, 200}, and therefore n(S) = 200

Let A represent the event that the integer chosen is divisible by 6, B the event that the integer chosen is divisible by 8, and A $\cup B$ the event that the integer chosen is divisible by both 6 and 8, i.e. by 24.

Then we need P(A $\cup B)$.

Now n(A) = $\left[\frac{200}{6}\right]$ = 33

n(B) = $\left[\frac{200}{8}\right]$ = 25, and

n(A $\cup B) = \left[\frac{200}{4}\right]$ = 8, where [x] stands for the highest integer in x

Hence P(A $\cup B)= P(A)+P(B)-P(A\cap B)$

$=\frac{33}{200}+\frac{25}{200}-\frac{8}{200}=\frac{50}{200}=\frac{1}{4}$

Example: A pair of dice are thrown. Find the probability of getting a total of either 5 or 11. The sample space has 36 outcomes when two are thrown.

***Solution*:** Let A be the event that a total of 5 occur and B be the event that a total of 11 occur. Then the two events are

A = {(1, 4), (2, 3), (3, 2), (4, 1)}, and

B = {(5, 6), (6, 5)}

Events A and B are mutually exclusive as a total of 5 and 11 cannot both occur together. Therefore,

$$P(A \cup B) = P(A) + P(B) = \frac{4}{36} + \frac{2}{36} = \frac{1}{6},$$

***Example*:** Three horses A, B and C are in a race. A is twice as likely to win as B, and B is twice as likely to win as C. What is the probability that A or B wins?

***Solution*:** Let $P(C) = p$

Then $P(B) = 2p$ as B is twice as likely to win as C.

Similarly $P(A) = 2P(B) = 4p$

Since A, B and C are mutually exclusive and collectively exhaustive, therefore the sum of their probabilities must be equal to 1. Thus,

$$p + 2p + 4p = 1 \; or \; p = \frac{1}{7}$$

$$P(A) = \frac{4}{7}, P(B) = \frac{2}{7} \text{ and } P(C) = \frac{1}{7}$$

Hence $P(A \cup B) = P(A) + P(B) = \frac{4}{7} + \frac{2}{7} = \frac{6}{7}$.

Theorem. If A, B and C are any three events in a sample space S, then the probability of at least one of them occurring is given by

$$P(A \cup B \cup C) = P(A) + P(B) + P(C) - P(A \cap B) - P(B \cap C) - P(A \cap C) + P(A \cap B \cap C)$$

Proof: Let us write $A \cup B \cup C = A \cup (B \cup C)$

$= A \cup D$, where $D = B \cup C$.

Then $P(A \cup B \cup C) = P(A \cup D)$

$= P(A) + P(D) - P(A \cap D)$

$= P(A) + P(B \cup C) - P\left[A \cap (B \cup C)\right]$

$= P(A) + P(B) + P(C) - P(B \cap C) - P\left[A \cap (B \cup C)\right]$

The distribution law of given

$A \cap B(B \cup C) = (A \cap B) \cup (A \cap C)$

Since the two set are not disjoint, their intersection is given $A \cap B \cap C$.

Therefore $P[A \cap (B \cup C)] = P[(A \cap B) \cup (A \cap C)$

$= P(A \cap B) + P(A \cap C) - P(A \cap B \cap C)$

Hence substitution of this result gives

$P(A \cap (B \cup C) = P(A) + P(B) + P(C) - P(A \cap C) - P(B \cap C) - P(A \cap C) + P(A \cap B \cap C)$

This result may be written as

$P(A_1 \cup A_2 \cup A_3) = \sum_i P(A_i) - \sum_{i<j} P(A_i \cap A_j) + P(A_1 \cap A_2 \cap A_3)$.

In general, the formula for k events is

$P(A_1 \cup A_2 \cup \ldots \cup A_k) = \sum_i P(A_i) - \sum_{i<j} P(A_i \cap A_j) + \sum_{i<j<l} P(A_i \cap A_j \cap_l)$

$-\ldots\ldots + (-1)^{k+1} P(A_1 \cap A_2 \cap \ldots \cap A_k)$.

Example: A card is drawn at random from a deck of ordinary playing cards. What is the probability that it is a diamond, a face card or a king?

Solution: Let A represent the event that the card drawn is a diamond, B, the event the card drawn is a face card, C the event that the card drawn is a king, A B, the event that the card drawn is both a diamond and face card, and so on. Then we need

$P(A \cup B \cup C) = P(A) + P(B) + P(C) - P(A \cap B) - P(B \cap C) - P(A \cap C) + P(A \cap B \cap C)$.

$$Now\, P(A) = \frac{n(A)}{n(S)} = \frac{13}{52}, (there\, are\, 13\, diamonds)$$

$$P(B)=\frac{n(B)}{n(S)}=\frac{12}{52},(\textit{there are 12 face card})$$

$$P(C)=\frac{n(C)}{n(S)}=\frac{4}{52},(\textit{there are 4 kings})$$

$$P(A\cap B)=\frac{n(A\cap B)}{n(S)}=\frac{3}{52},(\textit{diamond and face card})$$

$$P(B\cap C)=\frac{n(B\cap C)}{n(S)}=\frac{4}{52},(\textit{face card and kings})$$

$$P(A\cap C)=\frac{n(A\cap C)}{n(S)}=\frac{1}{52},(\textit{diamond and kings})$$

$$P(A\cap B\cap C)=\frac{n(A\cap B\cap C)}{n(S)}=\frac{1}{52},(\textit{diamond and face card and kings})$$

Hence, we get

$$P(A\cap B\cap C)=\frac{13}{52}+\frac{12}{52}+\frac{4}{52}-\frac{3}{52}-\frac{4}{52}-\frac{1}{52}+\frac{1}{52}=\frac{22}{52}=0.423$$

Conditional Probability

The sample space for an experiment must often be changed when some additional information pertaining to the outcomes of the experiment is received. The effect of such information is to reduce the sample space by excluding some outcomes as being impossible which before receiving the information were believed possible. The probabilities associated with such a reduce sample space are called conditional probabilities. The following example illustrates the concept of conditional probability.

Let us consider the die-throwing experiment with sample space S {1, 2, 3, 4, 5, 6}. Suppose we wish to know the probability of the outcomes, we are told that the die shows 6, say event A. If before seeing the outcomes, we are told that the die shows an even number of dots, say event B, then the information that the die shows an even number excludes the outcomes 1, 3 and 5, and thereby reduce the original sample space to a sample space that consist of 3 outcomes 2, 4 and 6, i.e. the reduced sample space is B = {2, 4, 6}. Then the desired probability in the reduced sample space B is since each outcomes in the reduced sample is equally likely. We call the conditional probability of the event A because it is computed under the condition that the die has shown even number of dots.

In other words,

$$P(die\,shows\,6 \quad die\,shows\,even\,number)\frac{1}{3},$$

Where the vertical line is read as "given that" and the information following the vertical line described the conditioning event. This is in face the probability of getting a 6 in the reduced sample space B, and is symbolized as P(A/ B). Thus we have the following:

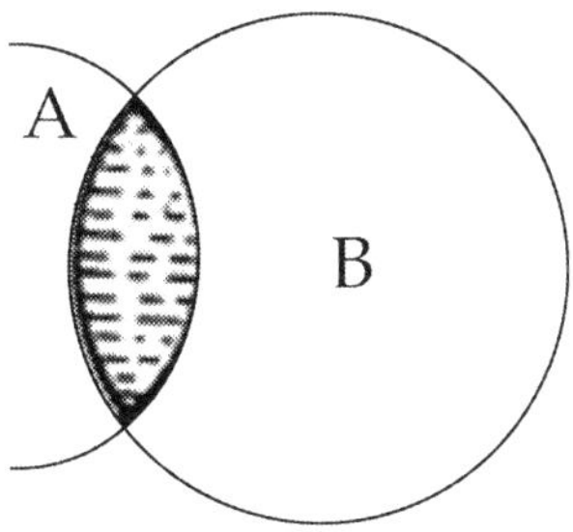

$$P(A / B) = \frac{number\ of\ sample\ space\ point\ in\ A \cap B}{number\ of\ sample\ points\ in\ B}$$

$$= \frac{n(A \cap B)}{n(B)}$$

Dividing the numerator and the denominator by the number of sample points in original sample space, i.e. n(S) we get

$$P(A/B) = \frac{n(A \cap B)}{n(B)} \cdot \frac{n(S)}{n(B)} = \frac{P(A \cap B)}{P(B)}$$

This illustration leads us to make the following definition of conditional probability.

If A and B are two events in a sample space S and if P(B) is not equal to zero, then the conditional probability of the event A given that event B has occurred, written as P(A/B), is defined by

$$P(A/B)=\frac{P(A\cap B)}{P(B)}$$

If P(B) = 0, then the conditional probability P(A/B) remains undefined

Similarly,
$$P(B/A)=\frac{P(A\cap B)}{P(B)}, where\, P(A)>0.$$

It should be noted that P (A/B) satisfies all the basic axioms probability, namely

(i) $0 \leq P(A/B)\leq 1.$

(ii) P(S/ B) = 1 [$\therefore S\cap B = B, \therefore P\left(\frac{S}{B}\right) = \frac{P(S\cap B)}{P(B)} = \frac{P(B)}{P(B)} = 1]$

(iii) P($A_1 \cup A_2\, B$) $= P(A_1 /$ B) + P(A_2 /B), provided the events $A_1\, and\, A_2$ are mutually exclusive.

Thus to determine the conditional probability P(A / B), either we directly calculate the probability of A relative to the reduced sample space B or we use P(A $\cap B)$ and P(B), the probabilities of the events in the original sample space S.

***Example*:** Two coins are tossed. What is the conditional probability that two head s result, given that there is at least one head?

***Solution*:** The sample space S for this experiment is

S = {HH, HT, TH, TT}.

Let A represent the event that two heads appear, and B, the event that there is at least one head. Then we need P(A/ B).

Since A = {HH}, B= {HH, HT, TH} and A $\cap B$ = {HH}

$$\therefore \quad P(A)=\frac{1}{4}, P(B)=\frac{3}{4} and\, P(A\cap B)=\frac{1}{4}$$

Hence P(A / B) = $\frac{P(A\cap B)}{P(B)} = \frac{1/4}{3/4} = \frac{1}{3}$

Example: A man tosses two fair dice. What is the conditional probability that the sum of the two dice will be 7, given that (i) the sum is odd, (ii) the sum is greater than 6, (iii) the two dice have the same outcomes?

Solution: The sample space S of this experiment consists of the following 36 equally likely outcomes;

S = { (1, 1), (1, 2), (1, 3), (1, 4), (1, 5), (1, 6)
(2, 1), (2, 2), (2, 3), (2, 4), (2, 5), (2, 6)
(3, 1), (3, 2), (3, 3), (3, 4), (3, 5), (3, 6)
(4, 1), (4, 2), (4, 3), (4, 4), (4, 5), (4, 6)
(5, 1), (5, 2), (5, 3), (5, 4), (5, 5), (5, 6)
(6, 1), (6, 2), (6, 3), (6, 4), (6, 5), (6, 6)}

Let A = {the sum is 7}

B = {the sum is odd}

C = {the sum is greater than 6}, and

D = {the two dice had the same outcomes}

Then,

A = {(1, 6), (2, 5), (3, 4), (4, 3), (5, 2), (6, 1)}

B = {(1, 2), (1, 4), (1, 6), (2, 1), (2, 3), (2, 5), (3, 2), (3, 4), ……, (6, 5)},

C = {(1, 6), (2, 5), (2, 6), (3, 4), (3, 5), (3, 6), (4, 3), (4, 4), ……, (6, 6)},

D = {(1, 1), (2, 2), (3, 3), (4, 4), (5, 5), (6, 6), (3, 2)},

$A \cap B$ = {(1, 6), (2, 5), (3, 4), (4, 3), (5, 2), (6, 1)},

$A \cap C$ = {(1, 6), (2, 5) (3, 4), (4, 3), (5, 2), (6, 1)}

$A \cap D = \varnothing$

$$\therefore P(A) = \frac{6}{36}, P(B) = \frac{18}{36}, P(C) = \frac{21}{36}, P(D) = \frac{6}{36}$$

$$P(A \cap B) = \frac{6}{36}, P(A \cap C) = \frac{6}{36} \text{ and } P(A \cap D) = 0.$$

Hence using the definition of conditional probability, we get

$$P(A/B) = \frac{P(A \cap B)}{p(B)} = \frac{6}{36} \times \frac{36}{18} = \frac{1}{3}$$

$$P(A/C) = \frac{P(A \cap C)}{pC} = \frac{6}{36} \times \frac{36}{21} = \frac{2}{7}$$

$$P(A/D) = \frac{P(A \cap D)}{p(D)} = 0 \times \frac{36}{6} = 0.$$

***Example*:** What is the probability that a randomly selected poker hand contains exactly 3 aces given that it contains at least 2 aces?

***4Solution*:** Let A represent the event that exactly 3 aces are selected and B, the even that at least 2 aces are selected. Then we need P(A / B).

Since a poker hand consists of 5 cards, therefore the sample space S contains

$$\binom{52}{5} = 2{,}598{,}960\, outcomes$$

Likewise n(A) = $\binom{4}{3}\binom{48}{2}$ outcomes;

n(B) = $\binom{4}{2}\binom{48}{3} + \binom{4}{3}\binom{48}{2} + \binom{4}{4}\binom{48}{1}$ because at least 2 aces means 2 or 3 or 4 aces; and

$$P(B) = \frac{\binom{4}{2}\binom{48}{3} + \binom{4}{3}\binom{48}{2} + \binom{4}{4}\binom{48}{1}}{\binom{52}{5}}$$

$$\text{Hence } P(A/B) = \frac{P(A \cap B)}{p(B)}$$

$$= \frac{\binom{4}{3}\binom{48}{2}}{\binom{4}{2}\binom{48}{3} + \binom{4}{3}\binom{48}{2} + \binom{4}{4}\binom{48}{1}}$$

$$=\frac{4.512}{108,336} = 0.0416$$

Multiplication Law

If A and B are any two events defined in a sample space S, then

$P(A \cap B) = P(A)P(B/A)$, provided P(A) $\neq 0$,

$= P(B)P(A/B)$, provided P(B) $\neq 0$,

Proof: The conditional probability of B given that A has occurred is

$$P(B/A) = \frac{P(A \cap B)}{p(A)}, \text{ where } P(A) \neq 0$$

Multiplying both side by P(A), we get

$P(A \cap B) = P(A).P(B/A)$

The second form is easily obtained by intersecting A and B. This is called the general rule of multiplication for probabilities.

Alternative Proof

Let S represent a sample space of an experiment having n equally likely outcomes (sample points). Let number of sample points contained in A (including those common to B); be the number of sample points in B and m_3 be the number of sample points belonging both to A and B. Then (assuming $m_1 > 0, m_2 > 0$),

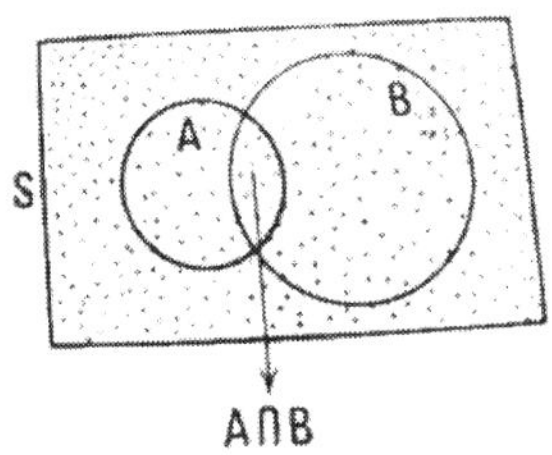

$$P(A \cap B) = \frac{m_3}{n}$$

The fraction $\frac{m_3}{n}$ may be written as $\frac{m_3}{n} = \frac{m_3}{m_1} \cdot \frac{m_1}{n}$

But $\frac{m_1}{n}$ = P(A) and $\frac{m_3}{n}$ = conditional probability of B, given that A has occurred, i.e. P(B/A)

Hence $P(A \cap B) = P(A)P(B/A)$

Since the joint event $A \cap B$ involves A and B symmetrically, therefore interchanging A and B we get

$P(A \cap B) = P(A)P(B/A)$.

This rule may be stated as below:

"The probability that two events involve A and B will both occur multiplied by the conditional probability that the other event will occur given that the first event has occurred."

Corollary

This rule may be extended to several events. In case of three events A, B and C, we have

$P(A \cap B \cap C) = P(D \cap C)$, *where* $D = A \cap B$

$= P(D)\ P(C/A \cap B)$

$= P(A)\ P(B/A)\ P(C/A \cap B)$;

Similarly, for more than three events, the formula may be proved by mathematical induction.

Example: A box consists of 15 items, 4 of which are defective and 11 are good. Two items are selected. What is the probability that the first is good and the second items is defective?

Solution: Let A represent the event that the first item selected is good and B, the event that the second item is defective.

Then we need to calculate the probability of the joint event $A \cap B$ by the rule $P(A \cap B) = P(A).\ P(B/A)$. Now $P(A) = \frac{11}{15}$

Given that event A has occurred, there remains 14 items of which 4 are defective. Therefore the probability of selecting a defective after good has been selected, i.e. P(B/A) $= \frac{4}{14}$.

Hence P(A $\cap B$) = P(A). P (B/A) = $\frac{11}{15} \times \frac{4}{14} = 0.16$

Example: Two cards are dealt from a pack of ordinary playing cards. Find the probability that the second card dealt is a heart.

Solution: Let H_1 represent the event that the first card dealt is a heart, and H_2, the event that the second dealt is a heart. Then

P(second card is heart) = P(first card is a heart and second card is a heart) + P(first card is not a heart and second card is a heart).

i.e. P(H_2) = P $(H_1 \cap H_2)$ + P $((H_1) P(H_2) P(H_1)$

$= \left(\frac{13}{52} \times \frac{12}{51}\right) + \left(\frac{39}{52} \times \frac{13}{51}\right)$

$= \frac{1}{17} + \frac{13}{68} = \frac{17}{68} = \frac{1}{4}$

Example: Box A contains 5 green and 7 red balls. Box B contains 3 green, 3 red and 6 yellow balls. A box is selected at random and a ball is drawn at random from it. What is the probability that the ball drawn is green?

Solution: Let E represent the event that the green ball is drawn. Then E can occur in one of the following two mutually exclusive ways:

(i) Box A is selected and a green ball is drawn, i.e. $A \cap E$, or

(ii) Box B is selected and a green ball a drawn, i.e. B $\cap E$.

Therefore, P(green ball) = P(box A and green ball) + P(box B and green ball)

= P(box A) P(green ball/ box A) + P(box B) P(green ball/ box B)

In symbols, P(E) = P(A $\cap E$) + P(B $\cap E$)

= P(A) P(E/A) + P(B)P(E/B)

$$= \frac{1}{2}.\frac{5}{2} + \frac{1}{2}.\frac{3}{12} = \frac{1}{3}$$

Example: An urn contains 10 white and 3 black balls. Another urn contains 3 white and 5 black balls. Two ball are transferred from first urn and placed in the second and then one ball is taken from the latter. What is the probability that it is a white ball?

Solution: Let A represent the event that 2 balls are drawn from the first urn and transferred to the second urn. Then A can occur in the following three mutually exclusive ways:

$A_1 = 2\, white\, balls$

$A_2 = 1\, white\, ball\, and\, 1\, black\, ball.$

$A_3 = 2\, black\, balls$

Thus, $P(A_1) = \binom{10}{2} + \binom{13}{2} = \frac{45}{78}$

$P(A_2) = \binom{10}{1}\binom{3}{1} + \binom{13}{2} = \frac{30}{78}$

$P(A_3) = \binom{3}{2} + \binom{13}{2} = \frac{3}{78}$

The second urn after having transferred 2 balls from the first turn now contains

(i) 5 white and 5 black balls (2 white balls transferred)

(ii) 4 White and 6 black balls (1 white and 1 black ball transferred)

(iii) 3 White and 7 black balls (2 black balls transferred)

Let W represent the event that a white ball is drawn from the second urn after having transferred 2 balls from the first urn.

Then

$P(W) = P(W \cap A_1) + P(W \cap A_2) + P(W \cap A_3)$

Now $P(W \cap A_1) = \frac{5}{10} \times \frac{45}{78} = \frac{15}{52}$

$$P(W \cap A_2) = \frac{4}{10} \times \frac{30}{78} = \frac{2}{13}, and$$

$$P(W \cap A_3) = \frac{3}{10} \times \frac{3}{78} = \frac{2}{260}$$

Hence the required probability is

$$P(W) = \frac{15}{52} + \frac{2}{13} + \frac{2}{260} = \frac{59}{140} = 0.4538$$

Example: A card is drawn at random from a deck of ordinary playing cards. What is the probability that it is a diamond? A face card? A king?

Solution: Let A = the card drawn is a diamond

B = the card drawn is a face card, and

C = the card drawn is a king

Then we need

$$P(A \cup B \cup C) = P(A) + P(B) + P(C) - P(A \cap B) - P(B \cap C) - P(A \cap C) + P(A \cap B \cap C)$$

Now $P(A) = \frac{13}{52}, P(B) = \frac{12}{52}, P(C) = \frac{4}{52}$

$$P(A \cap B) = P(A)P(B / A) = \frac{13}{52} \times \frac{3}{13} = \frac{3}{52}$$

$$P(B \cap C) = P(B)P(C / B) = \frac{12}{52} \times \frac{4}{12} = \frac{4}{52}$$

$$P(A \cap C) = P(A)P\left(\frac{C}{A}\right) = \frac{13}{52} \times \frac{1}{13} = \frac{1}{52}$$

$$P(A \cap B \cap C) = P(A)P(B / A)P(C / A \cap B) = \frac{13}{52} \times \frac{3}{13} \times \frac{1}{3} = \frac{1}{52}$$

Hence, we get

$$P(A \cup B \cup C) = \frac{13}{52} + \frac{12}{52} + \frac{4}{52} - \frac{3}{52} - \frac{4}{52} - \frac{1}{52} + \frac{1}{52}$$

$$= \frac{22}{52} = 0.423$$

Example: Three urns of the same appearance are given as follows:

Urn A contains 5 red and 7 white balls.

Urn B contains 4 red and 3 white balls.

Urn C contains 3 red and 4 white balls.

An urn is selected at random and a ball is drawn from the urn.

(i) What is the probability that the ball drawn is red?

(ii) If the ball drawn is red, what is probability that it came from urn A?

Solution: Here we first select one of the three urns and then we draw a ball which is either red (R) or white (W). In other words, we perform a sequence of two experiments. This process is described by the probability tree diagram, in which each branch of the tree gives the respective probability.

Now the probability of selecting urn A, for instance, and then a red ball (R) is $\frac{1}{3}, \frac{5}{12} = \frac{5}{36},$ because the probability that any particular path of the tree occurs is, by the multiplication law, the product of the probability of each branch of the path.

(i) Now the probability of drawing a red ball is given by the relation

P(R) = P(A) P(R/A) + P(B) P(R/B) + P(C) P(R/C)

As there are three mutually exclusive paths leading to the drawing of a red ball

Hence P(R) $= \frac{1}{3}.\frac{5}{12} + \frac{1}{3}.\frac{4}{7} + \frac{1}{3}.\frac{3}{7}$

$= \frac{119}{252} = 0.4722$

(ii) Here we need the probability that urn A is selected, given that the ball drawn is red, that is, P(A/R)

By definition P(A/R) = $\frac{P(A \cap R)}{P(R)}$

But $P(A \cap R)$ = probability that urn A is selected and a red ball is drawn

$$= \frac{1}{3} \times \frac{5}{12} = \frac{5}{36}$$

Hence $P(A/R) = \dfrac{5/36}{11/252}$

$$= \frac{35}{119} = 0.294$$

Independent and Dependent Events

Two events A and B in the same sample space S are defined to be **independent** (or statistically independent) *if the probability that one event occurs is not affected by whether the other event has or has not occurred.* That is,

Two events A and B in the same sample space S are defined to be **independent** (or statistically independent) *if the probability that one event occurs is not affected by whether the other event has or has not occurred.* That is,

$$P(A/B) = P(A) \, and \, P(B/A) = P(B)$$

It then follows that two events $A \, and \, B$ are independent if and only if

$$P(A \cap B) = P(A)P(B)$$

The events A and B are defined to be **dependent** if $P(A \cap B) \neq P(A) \times (B)$. This means that *the occurrence of one of the events in some way affect the probability of the occurrence of the other event.*

It is to be emphasized that two mutually exclusive events A and B are independent if and only if $P(A)P(B) = 0$, which is true when either $P(A) = 0 \, or \, P(B) = 0$. If both events A and B have non–zero probabilities, they must have a sample point in common. Thus two events that are independent, can never be mutually exclusive, are also dependent events, and two events that are non-mutually exclusive, may either be independent or dependent events.

Three events A, B and C, all defined on the same sample space, are said to be mutually independent if they satisfy the following conditions:

(i) They are pairwise independent, i.e.

$P(A \cap B) = P(A)P(B)$; $P(A \cap C) = P(A).P(C)$;

$P(B \cap C) = P(B).P(C)$

(ii) $P(A \cap B \cap C) = P(A).P(B).P(C)$

In general, the k events $A_1, A_2, \ldots, A_k$ are defined to be mutually independent if and only if the probability of the intersection of any 2, 3,…, or k of them equals the product of their respective probabilities.

Example: Two events A and B are such that P(A) = $\frac{1}{4}$. P(A/,B) = $\frac{1}{2}$ *and* P(B/A) = $\frac{2}{3}$

(i) Are A and B independent events?

(ii) Are A and B mutually exclusive events?

(iii) Find P(A $\cap B)$ and P(B)

Solution: (i) If A and B are independent events, then P(A/B) = P(A)

Now P(A) = $\frac{1}{4}$ *and* p(A/ B) = $\frac{1}{2}$ $i.e\, P(A/B) \neq P(A)$

Hence A and B are not independent events.

(ii) If A and B are mutually exclusive events, then P(A/B) = 0

But it is given that P(A/B) = $\frac{1}{2}$

Hence A and B are not mutually exclusive events

(iii) Now P(A $\cap B)$ = P(A) P(B/A) = $\frac{1}{4} x \frac{2}{3} = \frac{1}{6}$

By definition, we have

P(B) P(A/B) = P(A) P(B/A)

or P(B) $\left(\frac{1}{2}\right)=\left(\frac{1}{4}\right)\left(\frac{2}{3}\right) so that$ P(B) = $\frac{1}{4}x\frac{2}{3}x\frac{2}{1}=\frac{1}{3}$

***Example*:** Two fair dice, one red and one green, are thrown. Let A denote the event that the red die shows and even number and B, the event that the green die shows a 5 or a 6. Show that the events A and B are independent.

***Solution*:** The sample space S contains 36 equally likely outcomes when two dice are thrown.

Given A = event that red die shows an even number,

B = event that green die shows a 5 or a 6, and therefore

$A\cap B=$ event that red die shows an even number and green die shows a 5 or a 6

Then A contains 18 outcomes, B contains 12 and $A\cap B$ contains only 6 outcomes.

Associating with each outcome a probability of $\frac{1}{36}, we\ get$

P(A) = $\frac{18}{36}=\frac{1}{2}; P(B)=\frac{12}{36}=\frac{1}{3}$

$P(A\cap B) = \frac{6}{36}=\frac{1}{6}$

Since P($A\cap B)=\frac{1}{6}=\frac{1}{2}x\frac{1}{3}=P(A)P(B), therefore\ the\ events\ A\ and\ B\ are\ independent$.

Alternatively,

P(A /B) = $\frac{P(A\cap B)}{P(B)} = \frac{1/6}{1/2}=\frac{1}{3}=P(B)$.

Hence the events A and B are independent.

***Example*:** Let A be the event that a family has children of both sexes and B the event that a family has at most one boy. If a family is known to have (i) three children, then show that A and B are independent events, (ii) four children, then show that A and B are dependent events.

Solution: Let b denote a boy and g a girl. Then

(i) the equiprobable sample space S would be

S = {bbb, bbg, bgb, gbb, bgg, gbg, ggb, ggg}

That two events are

A = {children of both sexes}

= {bbg, bgb, gbb, bgg, gbg, ggb}, and

B = {at most one boy}

= {bgg, gbg, ggb, ggg}

The event $A \cap B$ is

$A \cap B = \{bgg, gbg, ggb\}$

Thus their respective probabilities are

P(A) = $\frac{6}{8} = \frac{3}{4}, P(B) = \frac{4}{8} = \frac{1}{2}, and\ P(A \cap B) = \frac{3}{8}$

Now P(A) P(B) = $\frac{3}{4} x \frac{1}{2} = \frac{3}{8} = P(A \cap B)$.

Hence A and B are independent.

(ii) the sample space S may be represented by the following 16 equally likely outcomes.

S= {bbbb, bbbg, bbgb, bbgg, bgbb, bgbg, bggb, bggg, gbbb, gbbg, gbgb, gbgg, ggbb, ggbg, gggb, gggg}

A = {bbbg, bbgb, bbgg, bgbb, bgbg, bggb, bggg, gbbb, gbbg, gbgb, gbgg, ggbb, ggbg, gggb}

B = {bggg, gbgg, ggbg, gggb} and

$A \cap B$ = {bggg, gbgg, ggbg, gggb}

Their probabilities are

P(A) = $\frac{14}{16} = \frac{7}{8}, P(B) = \frac{15}{16} and\ P(A \cap B) = \frac{4}{16} = \frac{1}{4}$

Now P(A) P(B) = $\frac{7}{8} . \frac{5}{16} \neq P(A \cap B)$.

Hence A and B are dependent events.

Theorem: If A and B are two independent events, then

P(A $\cap B$) = P(A) P(B)

Proof: Since A and B are independent events, therefore

P(A) = P(A/B) and P(B) = P(B/A)

Substituting these result in the general rule of multiplication, we obtain

P(A $\cap B)$ = P(A) P(B)

Theorem: If A and B are two independent events in a sample space S, then (i) A and B are independent, (ii) A and B are independent and (iii) A and B are independent

Proof. The events (A $\cap B$) and (A $\cap B$) are mutually exclusive and their union is A, i.e. A = (A $\cap B$) $\cup(A \cap B)$.

Therefore P(A) = P(A $\cap B$) + P(A $\cap \overline{B}$)

or (A $\cap \overline{B}$) = P(A) - P(A $\cap B$)

= P(A) – P(A) P(B) [$\therefore$ A and B are independent]

= P(A) [1- P(B)]

= P(A) P($\overline{B}$)

Hence A and B are independent

(ii) Similarly,

P(B) =) (B $\cap A) = P(B) -$ (B $\cap \overline{A}$)

or P(B $\cap \overline{A}) =$ P(B) – P(B $\cap A)$

= P(B)n- P(B)P(A) (A and B are independent)

= P(B) –[1 – P(A)] = P(B) P($\overline{A}$)

Therefore $\overline{A}$ and B are independent.

(iii) Using the Morgan Law, ($\overline{A} \cap \overline{B}$) $= \overline{A \cup B,}$ We have

P($\overline{A} \cap \overline{B}$) = P ($\overline{A \cup B,}$)

= 1 $-P(A \cup B)$

=1 $-P(A) - P(B) + P(A \cap B)$

$= 1 - P(A) - P(B) + P(A)P(B)$

$= [1 - P(A)][1 - P(B)] = P(A)P(B)$

Which shows that A and B are independent

Example: Two Cards are drawn from a well shuffled ordinary deck of 52 cards. Find the probability that they are both aces if the first card is (i) replaced, (ii) not replaced.

Solution: Let A denote the event ace on the first draw and B are independent.

(i) In case of replacement, event A and B are independent.

Thus P(both cards are aces) = $P(A \cap B) = P(A)P(B)$

$$\frac{4}{52} \times \frac{4}{52} = \frac{1}{169}$$

(ii) If the first card is not replaced, then A and B are dependent events and therefore

P(both card are aces) P(first card is an ace) x P(second card is an ace given that the first card is an ace)

i.e. $P(A \cup B) = P(A)P$ (B /A) = $\frac{4}{52} x \frac{3}{51} = \frac{1}{121}$

Example: A pair of fair dice is thrown twice. What is the probability of getting totals of 5 and 11?

Solution: Let A represent the event or getting a total 5 and B, the event of getting a total of 11 when two fair dice are thrown.

Then A can occur in the following two ways:

$A_1 = \{a\, total\, of\, 5\, occurs\, on\, the\, first\, throw\}$

$A_2 = \{a\, total\, of\, 5\, occurs\, on\, the\, second\, throw\}$

And B can occur in the following two ways:

$B_1 = \{a\, total\, of\, 11\, occurs\, on\, the\, first\, throw\}$

$B_2 = \{a\, total\, of\, 11\, occurs\, on\, the\, second\, throw\}$

The joint event $A \cap B$ can occur in two mutually exclusive ways $A_1 \cap B_2 \, or \, B_1 \cap A_2, i.e.$

$$A \cap B = (A_1 \cap B_2) \cup (B_1 \cap A_2)$$

Therefore $P(A \cap B) = P(A_1 \cap B_2) + P(B_1 \cap A_2)$

$= P(A_1)P(B_2) + P(B_1)P(A_2)$ (events are independent)

$$= \frac{4}{36} \cdot \frac{2}{36} + \frac{2}{36} \cdot \frac{4}{36} = \frac{1}{162} + \frac{1}{162} = \frac{1}{81}.$$

Bayes Theorem: If the events $A_1, A_2, \ldots, A_k$ form a partition of a sample space S, that is, the events A_i are mutually exclusive and their union is S, and if B is any other event of S such that it can occur only if one of the A_i occurs, then for any i,

$$P(A_i / B) = \frac{P(A_i)P(B / A_i)}{\sum_{i=1}^{k} P(A_i).P(B / A_i)}, \; for \, i = 1, 2, \ldots, k.$$

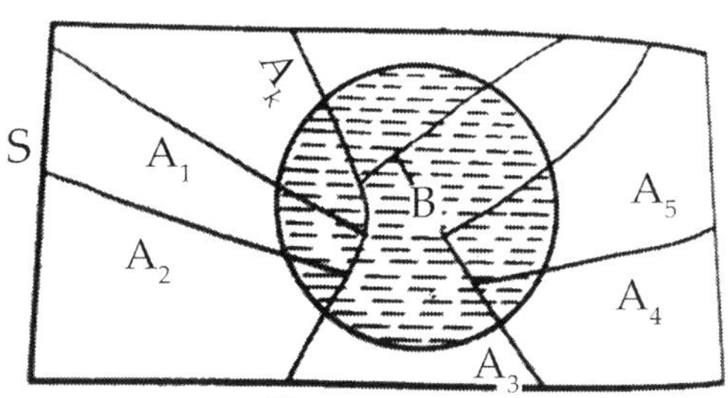

B is shaded

The result is known as Bayes' theorem after an English clergyman Thomas Bayes (1702- 1761) who derived it and first used in a paper that was published posthumously in 1763. It should be noted that the original probabilities $P(A_i)$ are known as the a *priori* probabilities and the conditional probabilities $P(A_i / B)$ are called the a posteriori or inverse probabilities because probabilities are revised after some additional information has been obtained. Bayes formula is also called the formula for probabilities of hypotheses on account of the reason that the events $A_1, A_2, \dots, A_k$ may be thought of as hypotheses to account for occurrence of the event B.

Example: In a bolt factory, machines $A, B\, and\, C$ manufacture $25, 35\, and\, 40$ percent of the total output, respectively. Of their outputs, $5, 4\, and\, 2$ percent, respectively, are defective bolts. A bolt is selected at random and found to be defective. What is the probability that the bolt came from machine A? B? C?

Solution: The a priori probabilities (before the information that the bolt is defective) are $P(A) = 0.25,\ P(B) = 0.35,$ and $P(C) = 0.40.$

Let E represents the event that a bolt is defective (D).

Then the conditional probabilities are

$$P(E/A) = 0.05, P(E/B) = 0.04, and\ P(E/C) = 0.02$$

The outcomes with their respective probabilities may be shown by a tree diagram as below:

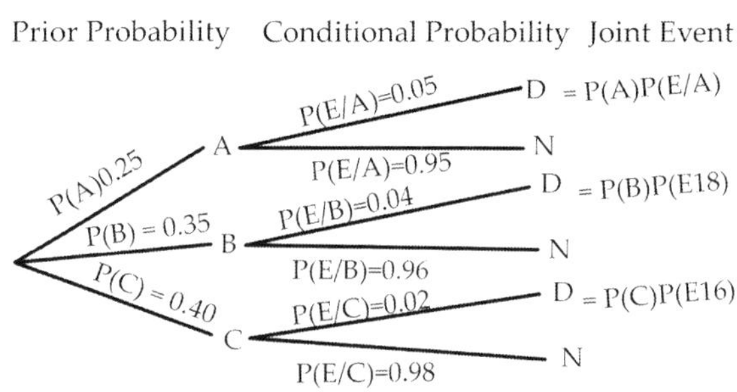

Now $P(A/E)$ is the a posteriori probability that the selected defective bolt came from machine Therefore, by Bayes theorem, we get

$$P(A/E) = \frac{P(A).P(E/A)}{P(A).P(E/A) + P(B).P(E/B) + P(C).P(E/C)}$$

$$=\frac{(0.25)(0.25)}{(0.25)(0.05)+(0.35)(0.04)+(0.40)(0.02)}$$

$$=\frac{0.0125}{0.0345}=0.362$$

Similarly, the posteriori probabilities of machines $B\, and\, C$ are $P(B/E)=0.406, P(C/E)=0.232$

SUMMARY AND DISCUSSION

In this chapter, we discussed some fundamental axioms of probability theory. The axioms or theorems which play very important role in probability theory. Also we discussed about conditional probability as a very important topic of probability theory.

Theorem if $\varnothing$ is the impossible event, then P($\varnothing$) = 0.

Law of Complementation

If $\bar{A}$ is the complement of an event A relative to the sample space S, then $P(\bar{A})=1-P(A)$.

Probability of Sub Event. If A and B are two events such that A $\subset B, then$ $P(A)\leq P(B)$.

Theorem: If A and B are any two events defined in a sample space S. Then

$$P(A\cap\bar{B})=P(A)-P(A\cap B)$$

Addition Law

This Law, often called the General Rule of Addition for probabilities may be stated as below: *"If two events A and B are not mutually exclusive, then the probabilities that at least one of them occurs, is given by the sum of the separate probabilities of events A and B minus the probability of the joint event A $\cap$B."*

If A and B are any two events defined in a sample space S, then

$$P(A\cup B)=P(A)+P(B)-P(A\cap B)$$

Corollary 1. If A and B are mutually exclusive events, then $P(A\cup B)=P(A)+P(B)$

Corollary 2. If $A_1, A_2, \ldots, A_k$ mutually exclusive events, then the probability that one of them occurs, is the sum of the probabilities of the separate events, i.e.

$$P(A_1\cup B_2\cup\ldots\ldots\cup A_k)=P(A_1)+P(A_2)+\ldots\ldots,+P(A_k).$$

Corollary 3. If A and B are any two events, then

$$P(A\cup B)\leq P(A)+P(B)$$

Conditional Probability

The sample space for an experiment must often be changed when some additional information pertaining to the outcomes of the experiment is received. The effect of such information is to reduce the sample space by excluding some outcomes as being impossible which before receiving the information were believed possible. The probabilities associated with such a reduce sample space are called conditional probabilities. The following example illustrates the concept of conditional probability.

$$P(A/B)=\frac{number\ of\ sample\ space\ point\ in\ A\cap B}{number\ of\ sample\ points\ in\ B}$$

$$=\frac{n(A\cap B)}{n(B)}$$

$$P(A/B)=\frac{P(A\cap B)}{P(B)}$$

It should be noted that P (A/B) satisfies all the basic axioms probability, namely

(i) $0\leq P(A/B)\leq 1.$

(ii) P(S/ B) = 1 [$\therefore S \cap B = B, \therefore P\left(\frac{S}{B}\right) = \frac{P(S \cap B)}{P(B)} = \frac{P(B)}{P(B)} = 1$]

(iii) P($A_1 \cup A_2$ B) $= P(A_1 /$ B) + P(A_2 /B), provided the events A_1 *and* A_2 are mutually exclusive.

The divide-and-conquer method. Here, the probabilities P(B) of various events B are obtained from conditional probabilities, where the Ai are appropriate events that figure a division of the sample space and have known probabilities $P(Ai)$. The probabilities $P(B)$ are then obtained by using the total probability theorem.

Finally, we have focused on a few side topics that strengthen our main themes. We have discussed the use of Bayes' rule in inference, which is an important application context. We have also discussed some basic principles of counting and combinatories, which are helpful in applying the counting method.

Multiplication Law

If A and B are any two events defined in a sample space S, Then

P(A $\cap B) = P(A)P(B/A)$, provided P(A) $\neq 0$,

$= P(B)P(A/B)$, provided P(B) $\neq 0$,

Independent and Dependent Events

Two events A and B in the same sample space S are defined to be independent (or statistically independent) if the probability that one event occurs is not affected by whether the other event has or has not occurred, that is

$$P(A\ \ B) = P(A) = P(A)P(B)$$

The event A and B are defined to be dependent if $P(A \cap B) \neq P(A) \times P(B)$. This means that the occurrence of one of the events in some way affect the probability of the occurrence of the other event.

It is to be emphasized that two mutually exclusive events A and B are independent if and only if $P(A).P(B)=0$, which is true when either $P(A)=0$ or $P(B)=0.$ If both events A and B have non–zero probabilities, they must have a sample point in common. Thus two events that are independent, can never be mutually exclusive, are also dependent events, and two events that are non-mutually exclusive, may either be independent or dependent events.

Three events A, B and C, all defined on the same sample space, are said to be mutually independent if they satisfy the following conditions:

(i) They are pairwise independent, i.e.

$P(A \cap B) = P(A)P(B)$

$P(A \cup C) = P(A)P(C);$

$P(B \cap C) = P(B)P(C)$

(ii) $P(A \cap B \cap C) = P(A)P(B)P(C)$

CHAPTER 3

RANDOM VARIABLE

INTRODUCTION

Usually, we are not interested in a particular outcome of a random experiment but our interest relates to some numerical description of the outcome. For example, when two coins are tossed, we may be interested only in the number of heads which appear and not in the actual sequence of heads and tails, which is not a numerical quantity. To express the outcomes in numbers, we assign to each non-numerical outcome of the sample space $S=\{HH, HT, TH, TT\}$, one of the numbers i (i = 0, 1, 2) corresponding to the number of heads appearing. That is, we express the outcomes in terms of numerical values as

Domain or sample space (E_i): $(HH),(HT),(TH),(TT)$

Range or corresponding value $X=F(E)$: 2,1,1,0

Again in the experiment of throwing a pair of dice, if we are interested only in the sum of the dots on the upper faces of the two dice and not in the particular dots. We assign to each possible outcome of the experiment, one of numbers i (i = 2, 3, 4, ... , 12) corresponding to the sum of dots appearing on their faces.

It is clear that the numbers 0, 1, 2 and 2, 3, 4, ... , 12 in the above cited examples, are random quantities determined by the outcomes of the random experiments. Such a numerical quantity whose value is determined by the outcomes of a random experiment, is called a random variable. Mathematically, we assign a single real number to each outcome of the sample space, and hence we state that a random variable is a real valued function defined on a sample space. Thus the number of heads obtained in tossing of two coins and the sum of the dots obtained with a pair of dice in the above examples are the values of random variables.

It should be noted that in the above definition, a function has been named a random variable. This terminology, though inappropriate and somewhat unfortunate, is universally accepted and used.

A random variable is also called a chance variable, a stochastic variable or simply a variate and is abbreviated as r.v. The random variables are usually denoted by capital letters such as while the values taken by them are represented by the corresponding small letters such as x, y, z. It is to be emphasized that more than one r.v. can be defined on the same sample space. There are two types of r.v's: the discrete and the continuous.

Every random trial results in two or more out comes and generally the attention is in a particular feature of the outcomes of the experiment, when a pair of dice is thrown, the interest may be in the sum of up turned dots on both the dice. In this experiment the total may be 2 (one on each die), 3(one on the first die and two on the other) and so it may be 4, 5, 6, 7, 8, 9, 10, 11 or, 12. These values connected with outcomes are the value of a random variable. In one more example total number of children in each of the fifty randomly chosen families. If no family has more than 5 children then the values of this random variable i.e., number of children in each family, would be 0, 1, 2, 3, 4, 5 i.e. no child, one child, two children, 3 children, 4 children and five children respectively.

Distribution Function

The distribution function of a random variable *X*, denoted by F(x), is defined by F(x) = P($x \leq x$). The Function F(x) gives the probability of the event that X takes a value less than or equal to specified value x. The cumulative distribution function (c. d. f) as it is the cumulative probability function of the X from the smallest up to specific value of x.

Since F (x) is probability, it is quite obvious that

$F(-\infty)$ = p $(\varnothing)$ = 0 and F $(+\infty)$ = p (S) = 1

Let a and b be two real numbers such that a $< b$. Then the probability of the interval (a, b) is $F(b)-F(a)=P(X\leq b)P(X\leq a)$

$= P(a<X\leq b)$,

Which is non- negative and hence F(x) is a non-decreasing function of x.

Again Lim *F(x+ h) = F(x),* is continuous on the right at each value of X.

A $d.f.F(x)$ thus has the following properties:

(i) $F(-\infty) = 0$, F $(+\infty) = 1$

(ii) F (x) is a non-decreasing function of x, i.e. F(x_1)$\leq F(x_2)$ if $x_1 \leq x_2$.

(iii) F (x) is continuous at least on the right of each x.

All random variable have distribution functions. Distribution functions for different $r.v's.$ are distinguished by using the notation $F_{x'}F_y^{'}$ etc.

Discrete Random Variable and Its Probability Distribution

A random variable X is defined to be discrete if it can assume values which are finite or countably infinite. (When X takes on finite numbers of various, they may be listed as $x_1, x_2, \ldots\ldots x_n,$ but in the countably infinite case, the values may be listed as $x_1, x_2, x_3, \ldots\ldots x_n, \ldots$ The numbers of heads obtained in coin tossing experiments, the number of defective items observed in a consignment, the number of fatal accidents, the number of bacteria in 1cs of water, etc., are the examples of discrete r,v.

Let X be a discrete r.v. taking on distinct values $x_1, x_2, \ldots, x_n, \ldots$ Then the function, denoted by p(x) or f(x), and defined by

$F(x_i) = P(X = x_i)$ for i = 1, 2 $\ldots, n, \ldots$

$= 0, \quad \text{for } x \neq x_i$

is called the probability function (p f) of the r. v. X takes the value x_i.

The distribution function for a discrete r.v. is F(x) = $\Sigma f(x_i)$,

Where sum is taken over all x_i that are less than or equal to x. Letting x tend to F (+ ∞), we have F (+$\infty) = \Sigma f(x_i) = 1$.

F (x) in case of a discrete r.v. is a step function. That is, its graph consist of horizontal line segments between any two successive values and has a step or jump of height $f(x_i)$, at each value of x_i. It should be noted that F(x) is continuous but between jumps, it is constant.

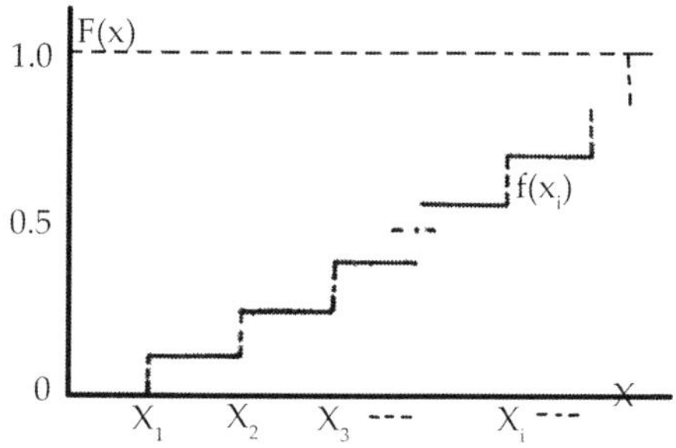

A discrete r.v may also be defined as a r.v. whose d.f jumps at the possible values of X and remains constant between adjacent jumps points. The height of a jump at each point x is the probability of the event X = x. That is, $f(x_i) = [\,jump\,at\,x_i\,] = P(X = x_i)$.

The set whose the elements are the ordered pairs $[x_i, f(x_i)], i = 1, 2 \ldots.$ defines the probability functions and probability distribution but they use them interchangeably, The probability distribution of a.r.v may conveniently be expressed either in a tabular from the showing all the possible values of X and the associated probabilities $f(x_i) = P(X - x_i)$ as

Value x_i	$x_1 \quad x_2 \quad \ldots. \quad x_n \quad \ldots.$

Probability $f(x_i)$	$f(x_1)$ $f(x_2)$..... $f(x_n)$

Or in the form of an equation for F (x) with a list of the possible values of X.

The graph of a probability distribution is obtained by locating the values $x_1, x_2, \ldots\ldots x_i,$

Along the horizontal axis and drawing vertical lines of heights equal to $f(x_1), f(x_2), \ldots\ldots, f(x_i), \ldots$ above them.

A probability distribution must satisfy the following two basic properties of probability

(i) $f(x_i) \geq 0$, for all i.

(ii) $\sum_i f(x_i) = 1$

In other words the probability of an outcome (x_i) is a greater than or equal to zero and sum of the probabilities associated with all possible outcomes must be unity.

Example: Find the probability distribution and distribution function for the number of heads when 3 balanced coins are tossed.

Solution: Construct a probability histogram and a graph of the distribution. The possible sample space for this experiment is

S = $\{HHH, HHT, HTH, THH, HTT, THT, TTH, TTT\}$.

Let X be the r.v that denotes the numbers of heads. Then the values of x are 0, 1, 2 and 3 and their probabilities are:

$$f(0) = P(X = 0) = P[\{TTT\}] = \frac{1}{8}$$

$$f(1) = P(X = 1) = P[\{HTT, THT, TTH\}] = \frac{3}{8}$$

$$f(2) = P(X = 2) = P[\{HHT\ HTH, THH\}] = \frac{3}{8}$$

$f(3) = P(X = 3) = P[\{HHH\}] = \frac{1}{8}$

Putting this information in the tabular from, we obtain the desired probability distribution of X as

Number of heads x_i	0	1	2	3
Probability $f(x_i)$	$\frac{1}{8}$	$\frac{3}{8}$	$\frac{3}{8}$	$\frac{1}{8}$

To obtain a formula, we need expressions for x heads to be selected out of 3 heads and for the denominators for all values. Now, x heads can be selected in $\binom{3}{x}$ ways total numbers of outcomes is 8. Therefore the desired probability distribution in the form of an equation is

$$f(x) = P(X = x) = \left(\frac{\frac{3}{x}}{8}\right) = \left(\frac{3}{x}\right)\left(\frac{1}{2}\right)^3 \text{ for } x = 0,1,2,3$$

For distribution function we compute the probabilities as below:

If $x < 0, we\, have\, P(X < x) = 0$

If 0 x $\leq x < 1, we\, have\, P(X < x) = P(X = 0) = \frac{1}{8}$

For 1 $\leq x < 2,$ we have

P(X $< x) = P(X = 0) + P(X = 1) = \frac{1}{8} + \frac{3}{8} = \frac{4}{8}$

Similarly, for 2 $\leq x < 3,$ we have

P (X $< x) = P(X = 0) + P(X = 1) P(X = 2) = \frac{4}{8} + \frac{3}{8} = \frac{7}{8}$

Finally for x ≥ 3, we have

$$P(X < x) = \sum_{i=0}^{3} P(X=i) = 1$$

Hence the desired distribution function is

$$F(x) = \begin{cases} 0, \; for\; x<0 \\ \frac{1}{8}, \; for\; 0<x<1 \\ \frac{4}{8}, \; for\; 1 \leq x<2 \\ \frac{7}{8}, \; 2 \leq x<3 \\ 1, \; for\; x \leq 3 \end{cases}$$

Example (a) Find the probability distribution of the sum of the dots when two fair dice are thrown (b) Use the probability distribution to find the probabilities of obtaining (i) a sum of 8 or 11. (ii) a sum that is greater than 8, (iii) a sum that is greater than 5 butt less than or equal to 10

Solution: (a) The sample space S for the experiment of throwing two dice contains 36 sample points, which are equally likely, i.e. each point has probability

Let X be random variable representing the sum of dots which appear on the dice. Then the values of the r.v. are 2, 3, 4,.., 12. The probabilities of these values are computed as below:

$$f(2) = P(X=2) = P\left[\{1,1\}\right] = \frac{1}{36},$$

as there is only one point resulting in a sum of 2;

$$f(3) = P(X=3) = P\left[\{(1,2),(2,1)\}\right] = \frac{2}{36}$$

$$f(4) = P(X=4) = P\left[\{(1,3),(2,2),(2,1)\}\right] = \frac{3}{36}$$

Similarly, $f(5) = \frac{4}{36}, f(6) = \frac{5}{36}, f(7) = \frac{6}{36}, f(8) = \frac{5}{36}, f(9) = \frac{4}{36},$

$f(10) = \frac{3}{36}, f(11) = \frac{2}{36}$ and $f(12) = \frac{1}{36}$

Therefore the desired probability distribution of the r.v. X is

X_i	2	3	4	5	6	7	8	9	10	11	12
$f(x_i)$	$\frac{1}{36}$	$\frac{2}{36}$	$\frac{3}{36}$	$\frac{4}{36}$	$\frac{5}{36}$	$\frac{6}{36}$	$\frac{5}{36}$	$\frac{4}{36}$	$\frac{3}{36}$	$\frac{2}{36}$	$\frac{1}{36}$

It is interesting the note that this result may also be expressed by the equation as

$$f(x) = \frac{6-|7-x|}{36}, \text{for } x = 2,3,4\ldots\ldots..$$

(b) Using the Probability distribution, we get the required probabilities as follows:

(i) p (a sum of 8 or 11) = P $[(X=8) or (X=11)]$

$= (X=8)+(X=11)$

$= f(8)+f(11) = \frac{5}{36} + \frac{2}{36} = \frac{7}{36}$

(ii) P (a sum that is greater than 8)

= P (X >8)

= P (X=9) + P (X = 10) P (X =11) + P (X =12)

$= f(9) + f(10)+f(11)+f(12)$

$= \frac{4}{36} + \frac{3}{36} + \frac{2}{36} + \frac{1}{36} = \frac{10}{36}$

(iii) P(a sum that is greater than 5 but less or equal to 10)

$= P(5 < X \leq 10)$

$= p(X=6)+P(X=7)+P(X=8)+P(X=9)+P(X=10)$

$$= f(6) + f(7) + f(8) + f(9) + f(10)$$

$$= \frac{5}{36} + \frac{6}{36} + \frac{4}{36} + \frac{3}{36} = \frac{23}{36}$$

CONTINUOUS RANDOM VARIABLE AND ITS PROBABILITY DENSITY FUNCTION

A random variable X is defined to be *continuous* if it can assume every possible value in an interval $[a,b]$, a $> b,$ where a and b may be $-\infty$ and $+\infty$ respectcvely. The height of a person, the temperature at a place, the amount of rainfall, time to failure for an electronic system, etc. are examples of continuous random variable.

A r.v. X may also be defined as continuous if its d.f. F(x) is continuous and is differentiable everywhere except at isolated points. In the given range. The graph of F(x) has no jumps or steps but is a continuous function for all x.

Let the derivative of F(x) be denoted by $f, i.e.$

$$\frac{d\,F(X)}{DX} = f(x)$$

Since F (x) is non – decreasing function of x, we have

(i) $f(x) \geq 0,$

$(ii) F(x) = \int f(x)dx, for\ al\ x$

The function $f(x)$ is called the probability density function, abbreviated to P.d.f. or simply density function of th r.v. X.

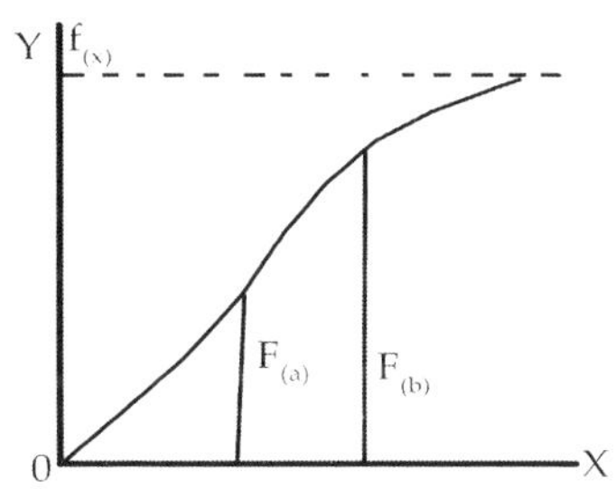

A p.d.f. has the following properties:

(i) $f(x) \geq 0,$ for all x

(ii) $\int_{-\infty}^{\infty} f(x)\,dx = 1$

(iii) The probability that X takes on a value in the interval $[c,d], c < d$ *is given by*

$P(c < x \leq d) = F(d) - F(c)$

$= \int_{-\infty}^{d} f(x)\,dx - \int_{-\infty}^{c} f(x)\,dx$

$= \int_{c}^{d} f(x)\,dx$

Which is the area under the curve y = $f(x)$ between X= c and X = d

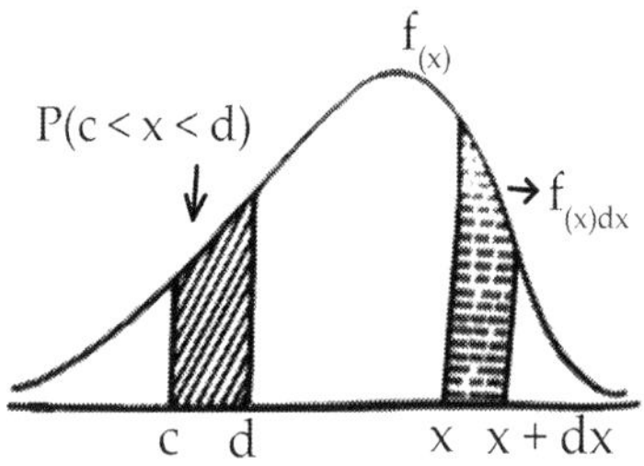

In other words, $f(x)$ is a non- negative function, the integration takes place over all possible values of r.v. X between the specified limits and the probabilities are given by appropriate areas under the curve.

Also, P(X $< X \leq x + dx) = F(x + dx) - F(x)$

$\cong f(x)\,dx$

The quantity $f(x)\,dx$ *is called* the probability differential or probability element (p.e.) of X.

Since P(X = κ) = $\int_{k}^{k} f(x)\,dx = 0,$

It should therefore be noted that the probability of a continuous r.v. X taking any particular value k is always zero. That is why probability for a continuous r.v is measureable only over a given interval.

Further, since for a continuous r.v. X, p(X = x) = 0 for every x, the following four probabilities are regarded the same:

P(c $\le X \le d), P(c <$ X ≤ d), P(c $\le X < d$ and $P($c $< X < d)$.

They may be different with a discrete r.v.

The value (expressed as in intervals) of a continuous r.v and their associated probabilities can be shown either in a tabular form or expressed by means of a formula.

Example: (a) Find the value of k so that the function $f(x)$ defined as follows, may be a density function $f(x) = kx,\ 0 \le x \le 2$

= 0, elsewhere

(b) Find also the probability that both of two sample values will exceed 1.

(c) Compute the distribution function F(x).

Solution: (a) The function f(x) will be a density function, if

(i) $f(x) \ge 0$ for every x, and

(ii) $\int_{-\infty}^{\infty} f(x)\,dx = 1$

The first condition is satisfied when k $\ge 0.The$ second condition with

be satisfied, if $\int_{-\infty}^{\infty} f(x)\,dx = 1,$

I.e. if 1 = $\int_{-\infty}^{0} f(x)\,dx + \int_{0}^{2} f(x)\,dx + \int_{2}^{\infty} f(x)\,dx$

i.e. if 1= $\int_{-\infty}^{0} 0\,dx \int_{0}^{2} kx\,dx + \int_{2}^{\infty} 0\,dx$

i.e. if $1 = 0 + \left[k\dfrac{x^2}{2}\right]_0^2 + 0 = 2k$

Thus, given $k = \dfrac{1}{2}$

Hence $f(x) = \begin{cases} \dfrac{x}{2}, for\ 0 \le x \le 2 \\ 0, elsewhere \end{cases}$

(b) $P(X > 1)$ = area of shaded region

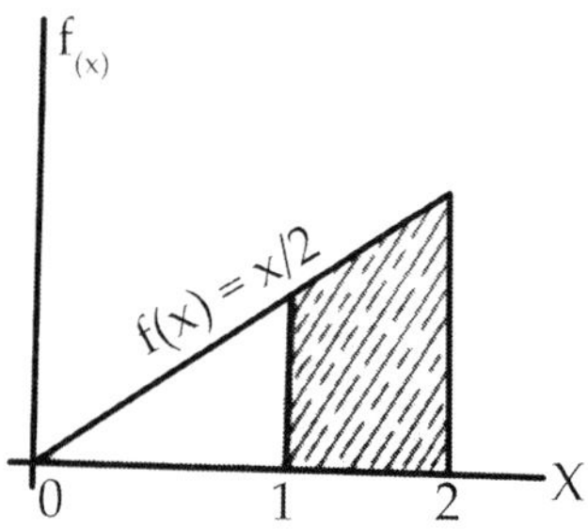

$= \int_1^2 f(x)\,dx$

$= \int_1^2 \dfrac{x}{2}\,dx = \left[\dfrac{x^2}{4}\right]_1^2 = \dfrac{3}{4}$

P (two sample values exceeding one) = $\dfrac{3}{4} \times \dfrac{3}{4} = \dfrac{9}{16}$

(c) To compute the distribution function, we find F(x)

$F(x) = P(X <) = \int_{-\infty}^{x} fat-(x)\,dx$

For any x such that $-\infty < x \le 0, F(x) = \int_{-\infty}^{x} 0\,dx = 0$

If 0 $< x \leq 2, we\, have\, F(x) = \int_{-\infty}^{0} 0\, dx + \int_{0}^{x} \left(\frac{x}{2}\right) dx = \left[\frac{x^2}{4}\right]_0^x = \frac{x^2}{4},$

and finally for x < 2 we have F (x) = $\int_{-\infty}^{0} 0\, dx \int_{0}^{2} \frac{x}{2}\, dx + \int_{0}^{2} 0\, dx = 1$

Hence F(x) = 0, for x < 0

= $\frac{x^2}{4},$ for 0 $\leq x \leq 2$

= 1, for x < 2

Example: A r.v. *X* is of continuous type with p.d.f.

$f(x) = 2x \quad 0 \leq x \leq 1,$

= 0, elsewhere.

Find (i) $P\left(X = \frac{1}{2}\right), (ii) P\left(X \leq \frac{1}{2}\right), (iii) P\left(X > \frac{1}{4}\right),$ (iv) $P\ (\frac{1}{4} \leq X < \frac{1}{2}),$ and

(v) $P\left(X \leq \frac{1}{2} \middle| \frac{1}{3} \leq X \leq \frac{2}{3}\right).$

Solution: Clearly $f(x) \leq 0\, and \int_{-\infty}^{\infty} f(x)\, dx = \int_{0}^{1} 2x\, dx = 1$

(i) Since f(x) is a continuous probability function therefore

P(X = $\frac{1}{2}) = 0.$

(ii) P(X $\geq \frac{1}{2}) = \int_{-\infty}^{0} 0\, dx + \int_{0}^{1/2} 2x\, dx = 0 + \left[x^2\right]_0^{1/2}$ $= \frac{1}{4}$

(iii) P(X $\geq \frac{1}{4}) = \int_{1/4}^{1/2} 2x\, dx = \left[x^2\right]_{1/4}$ $+ 0 = \frac{15}{16}$

(v) Applying the definition of conditional probability, we get

$$P\left(X \geq \frac{1}{2} \frac{1}{3} \geq X \geq \frac{2}{3}\right) = \frac{P\left(\frac{1}{3} \geq X \geq \frac{1}{2}\right)}{P\left(\frac{1}{3} \geq X \geq \frac{2}{3}\right)} = \frac{\int_{\frac{1}{3}}^{\frac{1}{2}} 2x dx}{\int_{\frac{1}{3}}^{\frac{2}{3}} 2x\, dx}$$

$$= \left[x^2\right]_{\frac{1}{3}}^{\frac{1}{2}} \div \left[x^2\right]_{\frac{1}{3}}^{\frac{2}{3}}$$

$$= \frac{5}{36} \times \frac{9}{3} = \frac{5}{12}.$$

JOINT DISTRIBUTIONS

The distribution of two or more random variable which are observed simultaneously when an experiment is performed, is called their joint distribution. It is n customary to call the distribution of a single r.v. as univariate. Likewise, a distribution involving two, three or many r.v simultaneously is referred to as bivariate, trivariate or multivariate.

Bivariate Distribution Function. Let X and Y be two r.v.s defined on the same sample space S. Then the function F(x, y) defined by *F(x, y)* = P(X $\leq x \, and \, Y \leq y$), where F(x,y) gives the probability that X will take on a value less to equal to x and, at the same time, Y will take on a value less than or equal to y, is called a bivariate or joint distribution function of X and Y.

(i) F (x - $\infty) = F(-\infty, y) = 0$, F(+ $\infty, +\infty$) = 1

(ii) F(x, y) is a non- decreasing function of x and y, and is continuous on the right

(iii) if $x_1 < x_2 \, and \, y_1 < y_2, then$

P($x_1 \leq X < x_2; y_1 \leq Y < y_2$ = p(X $< x_2, Y < y_2) - P(X \, x_2, Y < y_1 - P(X < x_1$, Y $< y_1$

= F($x_2 y_2)$ – F($x_2 \, y_1) - F(x_1 y_2) + F(x_1 y_1) > O$

The probability that a random point(X, Y) falls in the interval ($x_1 \le X < x_2$ $y_1 \le Y < y_2$)is shown graphically.

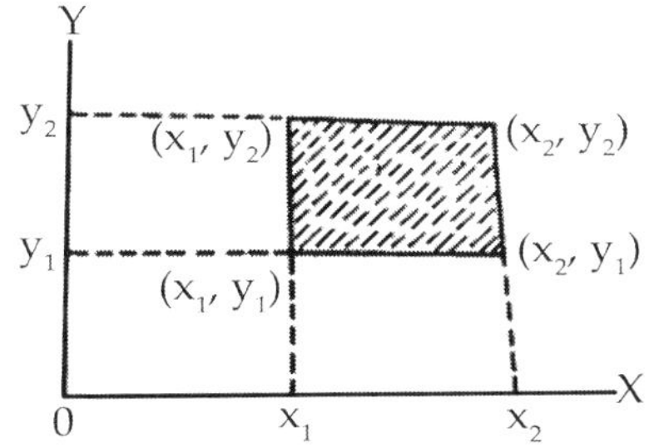

A bivariate distribution may be discrete when the possible values of (X, Y) are finite or countably infinite. It is continuous if (X, Y) can assume all values in some non-countable set of the plane. A bivariate distribution is said to be mixed when one r.v. is discrete and the other is continuous.

Bivariate Probability Function

Let X and Y be two discrete r .v defined on the some sample space S, X taking the values $x_1\, x_2, \ldots, X_m$ and Y is taking the values of $y_1\, y_2, \ldots y_n$ Then the probability that X takes on the value x_i and , at the same time Y takes on the value y_j,denoted by $f(x_i, y_j)$ Or P_{ij},*is* defined to be the joint probability function, also called bivariate probability function $f(x, y)$is a function whose value at the point (x_i, y_j) = P (X = $x_i and\, Y = y_j$),

i=1,2,……m., j=1,2,….,n.

The joint or bivariate probability distribution consisting of all pairs of values ($x_i y_j$)and their associated probabilities $f(x_i, x_{yj})$ *i.e.* the set of the triples [$x_i, y_j, f(x_i, y_j)$] can either be shown in a two-way table as on next page:

Joint Probability Distribution of X and Y

X / Y	y_1 y_1 y_1 y_1	$P(X = x_1)$
x_1 x_2 x_i x_m	$f(x_1, y_1)$ $f(x_1, y_2)$ ... $f(x_1, y_j)$....$f(x_1, y_n)$ $f(x_2, y_1)$ $f(x_2, y_2)$ ... $f(x_2, y_j)$....$f(x_2, y_n)$ $f(x_i, y_1)$ $f(x_i, y_2)$ ... $f(x_i, y_j)$....$f(x_i, y_n)$ $f(x_m, y_1)$ $f(x_m, y_2)$ ... $f(x_m, y_j)$....$f(x_m, y_n)$	$g(x_1)$ $g(x_2)$ $g(x_i)$ $g(x_m)$
$P(Y = y_j)$	h(y_1) h(y_2) h(y_j) h(y_n)	1

or be expressed by means of a formula for $f(x, y)$. The probabilities f(x, y) can be expressed by means of a formula for $f(x, y)$. The probabilities $f(x, y)$ can be obtained by substituting Appropriate values of x and y in the table or formula. A joint probability function has the following properties:

(i) $f(x_i, y_j) > 0$, for all (x_i, y_j), i.e. for I = 1,2, …..,m; j = 1,2, ….., n.

(ii) $\sum_i \sum_j f(x_i, y_j) = 1$

Marginal Probability Function

From the joint probability function for (X, Y), we can obtain the individual probability function of X and Y. Such individual probability functions are called marginal probability functions.

Let $f(x, y) to$ be the joint probability function of two discrete r.v.s X and Y. Then the marginal probability function of X is defined as

$$G(x_i)=\sum_{j=1}^{n} f(x_i, y_i)$$

$= f(x_i, y_1)+f(x_i, y_2) + \ldots.. + f(x_i, y_n)$ as x_i must occur

either with $y_1\ or\ y_2\ or \ldots or\ y_n$.

=P (X= x_i);

That is, the individual probability functions of X is found by adding over the rows of two-way table.

Similarly, the marginal probability function for Y is obtained by adding over the columns as

$$H(y_j) = \sum_{i=1}^{m} f(x_i, y_j) = P(Y = y_j)$$

The value of the marginal probabilities are often written in the margins of the joint tables as the row and column totals in the table. The probabilities in each marginal probability function add to 1.

Conditional Probability Function

Let X and Y be discrete r.v. with joint probability function $f(x, y)$. Then the Conditional probability function for X given Y = y, denoted as $f(x|y)$ is defined by

$$f((x_i|y_j) = \text{P (X} = x_i\ |Y = y_j\ )$$

$$= \frac{\text{P}(\text{X} = x_i\ and\ Y = y_j)}{P(Y = yx_j)}$$

$$= \frac{f(x_i, y_j}{g(x_i)},$$ where g(x) =>0.

Independence

Two discrete r.v. X and Y are said to be statistically independent, if and only if, for all possible pairs of values ($x_i, y_j)$ by the joint probability function $f(x,y)$ can be expressed as the product of the two marginal probability functions. That is, X and Y are independent, if

$$f(x,y) = P\left(X = x_i \, and \, Y = y_j\right)$$

= P(X = x_i) and P(Y = y_j) for all *i* and *j*.

It should be noted that the joint p.f. of X and Y when they are independent can be obtained by multiplying together their marginal probability functions.

Example: An urn contains 3 black, 2 red and 3 green balls, and 2 balls are selected at random from it. If X is the number of black balls and Y is the number of red balls selected, then find

(i) the joint probability function $f(x,y)$;

(ii) P (X + Y ≤ 1);

(iii) the marginal P. d f(x) and h(y);

(iv) the conditional p. d. $f(x|1)$;

(v) P(X= 0 $|Y = 1$), and

(vi) are X and Y independent

Solution: (i) The sample space S for this experiment contains $\left(\frac{8}{2}\right) = 28\,sample$ points. The possible values of X are 0, 1 and 2, and those for Y are 0, 1 and 2. The values that (X, Y) can take are (0, 0), (0, 1), (1, 0), (1, 1), (0, 2) and (2, 0) . We desire to find $f(x,y)$ for each values

(x, y)

Now $f(0,0) = P(X = 0a$ nd Y = 0), where the event (X = 0 and Y = 0) represents that neither black nor red ball is selected, implying that the 2 selected are green balls. This event therefore contains

$$\binom{3}{0}\binom{2}{0}\binom{3}{2} = 3\,sample\ points, and$$

$$f(0,0) = P(X = 0\,and\,Y = 0) = \frac{3}{28}$$

Again $f(0,1) = P(X = 0\,and\,Y = 1)$

= P (none is black, 1 is red and 1 is green)

$$\frac{\binom{3}{0}\binom{2}{1}\binom{3}{1}}{28} = \frac{6}{28}$$

Similarly, $f(1,1) = P(X = 1\,and\,Y = 1)$

= P (1 is black, 1 is red and none is green)

$$= \frac{\binom{3}{1}\binom{2}{1}\binom{3}{0}}{28} = \frac{6}{28}$$

Similarly, calculation gives the probabilities of other values and the joint p.f of X and Y is given as

(x,y)	(0, 0)	(0, 1)	(1, 0)	(1, 1)	(0, 2)	(2, 0)
$f(x,y)$	$\frac{3}{28}$	$\frac{6}{28}$	$\frac{9}{28}$	$\frac{6}{28}$	$\frac{1}{28}$	$\frac{3}{28}$

These probabilities can also be represented in another tabular form as follows:

Y X	0 1 2	$P(X = x_i)$ g(x)

0 1 2	$\frac{3}{28}$ $\frac{6}{28}$ $\frac{1}{28}$ $\frac{9}{28}$ $\frac{6}{28}$ $\frac{1}{28}$ $\frac{3}{28}$ 0 0	
$P(Y = y_j)$ h(y)	15 / 28 12 / 28 1/ 28	1

Clearly, this joint p.d. of the two r.v. (X, Y) can be represented by the formula

$$f(x,y) = \frac{\binom{3}{x}\binom{2}{y}\binom{3}{2-x-y}}{28}$$

x = 0, 1, 2

y = 0, 1, 2

$0 \leq x + y \leq 2.$

(ii) To compute $P(X + Y \leq 1)$, we see that $x + y \leq 1$ for the cells (0, 0), (0, 1) and 1, 0). Therefore,

$P(X + Y \leq 1) = f(0,0) + f(0,1) + f(1,0)$

$= \frac{3}{28} + \frac{6}{28} + \frac{9}{28} = \frac{18}{28} = \frac{9}{14}$

(iii) The marginal p.d. are:

X	0 1 2
g(x)	10/28 15/28 1/28

Y	0 1 2
g(x)	15/28 12/28 1/28

(iv) By definition the conditional p. d., $f(x|Y=1)$

$f(x|1)=P\ (X=x|Y=1)$

$$= \frac{P(X = x \text{ and } Y = 1)}{P(Y=1)}$$

Now h(1) = $x=\sum_{x=0}^{2} f(x,1)=\frac{6}{28}+\frac{6}{28}+0\frac{12}{28}=\frac{3}{7}$

Therefore, $f(x|1)=\frac{f(x,1)}{(h(1)}=\frac{7}{3}\ f(x,1), x=0,1,2$

That is, $f(0|1)=\frac{7}{3}f(0,1)=\frac{7}{3}f(0,1)=\left(\frac{7}{3}\right)\left(\frac{6}{28}\right)=\frac{1}{2}$

$$f(1|1)=\frac{7}{3}f(1,1)=\left(\frac{7}{3}\right)\left(\frac{6}{28}\right)=\frac{1}{2}$$

$$f(2|1)=\frac{7}{3}f(2,1)=\left(\frac{7}{3}\right)(0)=0$$

Hence the conditional p.d. of X given that Y = 1, is

X	0 1 2
$f(x\|1)$	1/ 2 1/2 0

(v). Finally, P(X=0 $\Big|\ Y=1)=f\ (0\Big|1=\frac{1}{2})$

(vi). We find that $f(0,1)=\frac{6}{28}$,

g(0) = $\sum_{y=2}^{2} f(0,y)=\frac{3}{28}+\frac{6}{28}+\frac{1}{28}=\frac{10}{28}$

h(1) = $\sum_{y=0}^{2} f(x,1) = \frac{6}{28} + \frac{6}{28} + 0 = \frac{12}{28}$

Now $\frac{6}{28} \neq \frac{10}{28} x \frac{12}{28}$,

i.e. $f(0, 1) \neq g(0)h(1)$.

and therefore, X and Y are not statistically independent.

Example: The joint p.d. of two discrete r.v. X and Y is given by

$f(x,y) = \frac{xy^2}{30}$ for x = 1, 2, 3 and y = 1, 2

Are X and Y independent?

Solution: The marginal p.d. for X is

g(x) = $\sum_{y} f(x,y)$

= $\sum_{y=1}^{2} f(x,y)$

= $\sum_{y=1}^{2} \frac{xy^2}{30} = \frac{x(1)^2}{30} + \frac{x(2)^2}{30} = \frac{x}{6}$, for x = 1. 2, 3;

And the marginal p.d. for Y is

h(y) = $\sum_{x} f(x,y)$

= $\sum_{x=1}^{3} \frac{xy^2}{30} = \frac{1y^2}{30} + \frac{2y^2}{30} + \frac{3y^2}{30} = \frac{y^2}{5}$, for = 1. 2

Clearly, $\frac{xy^2}{30} = \frac{x}{6} x \frac{y^2}{5}$. *for* $x = 1,2,3$ *and* $y = 1,2$,

i.e. $f(x,y) = g(x).h(y)$

Hence X and Y are independent

Continuous Bivariate Distribution

The bivariate probability density function of continuous r.v. X and Y is an integrable function $f(x,y)$ satisfying the following properties:

(i) $f(x,y) \geq o \; for\, all (x,y)$

(ii) $\int_{-\infty}^{\infty}\int_{-\infty}^{\infty} f(x,y)\,dx\,dy = 1$, and

(iii) P(a $\leq X \leq b, c \leq Y \leq d) = \int_{a}^{b}\int_{c}^{d} f(x,y)\,dy\,dx$.

The distribution function (d, f) of the bivariate r.v (X, Y) is defined by

F (x,y) = $\int_{-\infty}^{x}\int_{-\infty}^{y} f(u,v)\,dv\,du.$

It should be noted that analogous to the relationship $\frac{d}{dx}F(x) = f(x),$

we have $\frac{\delta^2 F(x,y)}{\delta x \delta y} = f(x,y)$ where F is differentable

The marginal p.d.f of the continuous r.v.X is

$$g(x) = \int_{-\infty}^{\infty} f(x,y)\,dy$$

And that of the r. v. Y is

h(y) = $\int_{-\infty}^{\infty} f(x,y)\,dx$

That is the marginal p.d.f. of any of the variables is obtained by integrating out the other variable from the joint p.d.f. between the limits -∞ *and* $+\infty$.

The conditional p.d.f. of the continuous r.v. X give that Y takes the value y, is defined to be $f(x|y) = \frac{f(x,y)}{\text{h}(\text{y})}$,

Where $f(x, y)$ and $h(y)$ are respectively the joint p.d.f. of X and Y, and the marginal p.d.f. of Y, and $h(y) > 0$.

Similarly, the conditional p.d.f. of the continuous r.v. Y given that X = x, is

$$f(y|x) = \frac{f(x,y)}{g(x)}$$, proved that $g(x) > 0$.

It is worth noting that the conditional p.d.f. satisfy all the requirements for a univariate density function.

Finally, two continuous r.v. X and Y are said to statistically independent, if and only if their joint density $f(x,y)$ can be factored is the form $f(x, y) = g(x). h(y)$ for all possible values of X and Y.

Example: Given the following joint p.d.f.

$$f(x,y) = \frac{1}{8}(6-x-y), 0 \le x \le 2; 2 \le y \le 4,$$

$$= 0, \text{ elsewhere}$$

(a) Verify that f(x,y) is a joint density function.

(b) Calculate (i) P$\left(X \le \frac{3}{2}, Y \le \frac{5}{2}\right), (ii) P(X+Y<3).$

(c) Find the marginal p.d.f. g(x) and h(y).

(d) Find the conditional p.d.f. $f(x|y)$ and $f(y|x)$

Solution: (a) The joint density f(x, y) will be a p.d.f. if

(i) f(x, y) ≥ 0 and (ii) $\int_{-\infty}^{\infty}\int_{-\infty}^{\infty} f(x,y)\,dx\,dy = 1$

Now $f(x,y)$ is clearly ≥ 0 for all x and y in the given region, and

$$\int_{-\infty}^{\infty}\int_{-\infty}^{\infty} f(x,y)\,dx\,dy = \frac{1}{8}\int_0^2\int_2^4 (6-x-y)\,dy\,dx$$

$$= \frac{1}{8}\int_0^2 \left[6y - xy - \frac{y^2}{2}\right]_2^4 dx$$

$= \frac{1}{8}\int_0^2 =(6-2x)\,dx = \frac{1}{8}\left[6x - x^2\right]_0^2$

= 1/8 [12 – 4] = 1

Thus $f(x, y)$ has properties of a joint p.d.f.

(b) (i) To determine the probability of values of the r.v (X,Y) falling in the region X $\leq \frac{3}{2}, Y \leq \frac{5}{2}$, we find

P(X $\leq \frac{3}{2}, Y \leq \frac{5}{2}) = \int_{x=0}^{\frac{3}{2}}\int_{y=2} \frac{1}{8}(6-x-y)\,dy\,dx$

$= \frac{1}{8}\int_0^{\frac{3}{2}}\left[6y - xy - \frac{y^2}{2}\right]_0^{\frac{5}{2}} dx$

$= \frac{1}{8}\int_0^{3/2}\left(\frac{15}{8} - \frac{x}{2}\right)dx = \frac{1}{64}\left[15x - 2x^2\right]_0^{\frac{3}{2}} = \frac{9}{32}$

(ii) P (X + Y $< 3) = \frac{1}{8}\int_0^1\int_2^{3-x}(6-x-y)\,dy\,dx \qquad (\therefore x+y \leq 3, \therefore y = 3-x)$

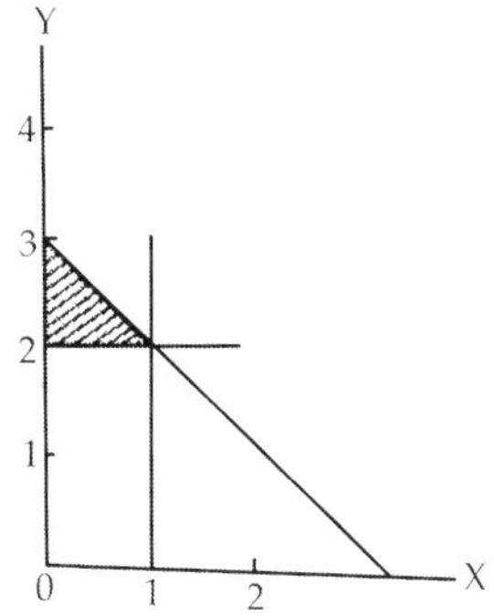

Event "X + Y ≤ 3" is shaded

$= \frac{1}{8}\int_0^1\left[6x - xy - \frac{y^2}{2}\right]_2^{3-x} dx$

$= \frac{1}{8}\int_0^1 \left(\frac{x^2}{2} - 4x + \frac{7}{2}\right)dx$

$= \frac{1}{8}\left[\frac{x^3}{6} - 2x^2 + \frac{7x}{2}\right]_1^2$

$= \frac{1}{8} x \frac{10}{6} = 5/24$

(c) The marginal p.d.f. of X is

$G(x) = \int_{-\infty}^{\infty} f(x,y)\,dy, \qquad -\infty < x < \infty$

$= \frac{1}{8}\int_2^4 (6 - x - y)\,dy, \qquad 0 \le x \le 2.$

$= \frac{1}{8}\left[6y - xy - \frac{y^2}{2}\right]_2^4 \qquad 0 \le x \le 2$

$= \frac{1}{4}(3 - x) \qquad 0 \le x \le 2$

$= 0, \qquad x < or\ x \ge 2.$

Similarly, the marginal p.d.f. of Y is

$H(y) = \frac{1}{8}\int_0^2 (6 - x - y)\,dx \quad 2 \le y \le 4$

$= \frac{1}{4}(5 - y) \qquad 2 \le y \le 4$

= 0, elsewhere.

(d) The conditional p . d . f of X given Y = y., is

$f(x|Y) = f\frac{(x,y)}{h(y)}, where\, h(y) > 0$

$$= \frac{\left(\frac{1}{8}\right)(6-x-y)}{(1/4)(5-y)} = \frac{6-x-y}{2(5-y)}$$

And the conditional p.d.f. of Y given X = x, is

$$f(y|x) = \frac{f(x,y)}{h(y)}, \text{ where } g(x) > 0$$

$$= \frac{(1/8)(6-x-y)}{(1/4)(3-x)} = \frac{6-x-y}{2(3-x)},$$

Example: Let the bivariate continuous random variable (X, Y) have the joint probability density functions given by

$$f(x,y) = x^2 + \frac{xy}{3}, 0 \le x \le 1, 0 \le y \le 2,$$

Check hat $f(x,y)$ is a p.d.f.

Find the marginal p.d.f.

Find the conditional p.d.f. and verify that $f(y|x)$ is p.d.f.

Solution: (a) The function $f(x,y)$ *will be* a p.d.f. if

(i) $f(x,y) \ge 0$ *and* (ii) $\int_{-\infty}^{+\infty}\int_{-\infty}^{+\infty} (x,y)\,dx\,dy = 1.$

Now $f(x,y)$ is clearly > 0 for all x and y in the given interval and

$$\int_{-\infty}^{+\infty}\int_{-\infty}^{+\infty} f(x,y)\,dx\,dy = \int_0^2\int_0^1 (x^2 + \frac{xy}{3}\,dx\,dy$$

$$= \int_0^2 \left[\frac{x^3}{3} + \frac{x^2y}{6}\right] dy = \int_0^2 \left(\frac{1}{3} + \frac{y}{6}\right) dy$$

$$= \left[\frac{y}{3} + \frac{y^2}{12}\right]_0^2 = \frac{2}{3} + \frac{4}{12} = 1$$

(b) The marginal p.d.f. are

$$g(x) = \int_{-\infty}^{\infty} f(x,y)\,dy = \int_0^2 \left(x^2 + \frac{xy}{3}\right) dy$$

$$= \left[x^2 y + \frac{x^2 y}{6}\right]_0^2 = 2x^2 + \frac{2}{3}x = \frac{2}{3}x(1+3x); 0 \le x \le 1$$

$$H(y) = \int_{-\infty}^{\infty} f(x,y)\,dx = \int_0^1 \left(x^2 + \frac{xy}{3}\right) dx$$

$$= \left[\frac{x^3}{3} + \frac{x^2 y}{6}\right]_0^1 = \frac{1}{3} + \frac{y}{6} = \frac{1}{6}(2+y) \qquad 0 \le y \le 2$$

(c) The conditional p.d.f. of Y for given $X = x$ is

$$F(y/x) = \frac{f(x,y)}{h(y)} = \frac{x^2 + \frac{xy}{3}}{\frac{1}{6}(2+y)} = \frac{6x^2 + 2xy}{2+y}, 0 \le x \le 1, \le 1, 0 \le y \le 2$$

And the conditional p.d.f. of Y for given X = x is

$$F(y/x) = \frac{f(x,y)}{g(x)} = \frac{x^2 + \frac{xy}{3}}{\frac{2}{3}x(1+3x)} = \frac{3x^2 + xy}{6x^2 + 2x} = \frac{3x+y}{6x+2}, \quad 0 \le y \le 2, 0 \le x \le 1$$

To verify that the conditional p.d.f. $F(x/y)$ is a p.d.f. we have

$$\int_0^1 \frac{6x^2 + 2xy}{2+y}\,dx = \frac{1}{2+y}\left(2x^3 + x^2 y\right)\Big|_0^1$$

$$= \frac{1}{2+y}(2+y) = 1, \quad \text{for all y.}$$

Hence the conditional p.d.f. $F(x/y)$ satisfies the requirements for a unvariate density function.

SUMMARY AND DISCUSSION

In this chapter, we discussed some fundamental concepts of random variable, probability distribution, discrete distribution, continuous distribution joint probability distribution and marginal probability distribution. Also we discussed important properties and explained how these distribution can be used in different types of real life problems.

Random Variable: A variable whose values depend upon the outcomes of the random experiment is called a random variable. We will denote the random by the capital letters X, Y or Z and their value by the corresponding small letters, x, y or z.

DISTRIBUTION FUNCTION

The distribution function of a random variable X, denoted by F(x), is defined by F(x) = P($X \leq x$). The Function F(x) gives the probability of the event that X takes a value less than or equal to a specified value x. The cumulative distribution function (cdf) as it is the cumulative probability function of the X from the smallest up to specific value of x.

Since F(x) is probability, it is quite obvious that

$F(-\infty) = P(\varnothing) = 0$ and $F(+\infty) = p(S) = 1$

DISCRETE RANDOM VARIABLE AND ITS PROBABILITY DISTRIBUTION

A random variable X is defined to be discrete if it can assume values which are finite or countably infinite. Let X be a discrete r.v. taking on distinct values $x_1, x_2, \ldots, x_n, \ldots$

Then the function, denoted by p(x) or f(x), and defined by

$F(x_i) = P(X = x_i)$ for $i = 1, 2 \ldots, n, \ldots$

$= 0,$ for $x \neq x_i$

is called the probability function (p.f) of the r.v. X takes the value x_i.

The distribution function for a discrete r.v. is

$$f(x)=\sum_i f(x_i)$$

Where sum is taken over all x_i that are less than or equal to x.

Letting x tend to + ∞, we have F(+ $\infty) = \Sigma f(x_i) = 1$.

F(x) in case of a discrete r.v. is a step function. That is, its graph consist of horizontal line segments between any two successive values and has a step or jump of height $f(x_i)$, at each value of x_i. It should be noted that F(x) is continuous but between jumps, it is constant.

A probability distribution must satisfy the following two basic properties of probability

(i) $f(x_i) \geq 0$, for all i.

(ii) $\sum_i f(x_i) = 1$

CONTINUOUS RANDOM VARIABLE AND ITS PROBABILITY DENSITY FUNCTION

A random variable X is defined to be *continuous* if it can assume every possible value in an interval in an $[a,b]$, a $> b$, where a and b may be $-\infty$ and $+\infty$ respectcvely. The height of a person, the temperature at a place, the amount of rainfall, time to failure for an electronic system, etc. are examples of continuous random variable.

A r.v. X may also be defined as continuous if its d.f. F(x) is continuous and is differentiable everywhere except at isolated points. In the given range. The graph of F(x) has no jumps or steps but is a continuous function for all x

Let the derivative of F(x) be denoted by f, $i.e.$

$$\frac{dF(X)}{DX} = f(x)$$

Since F (x) is non – decreasing function of x, we have

(i) $f(x) \geq 0, (ii)\ F(x) = \int f(x)dx, for\ al\ x$.

The function $f(x)$ is called the probability density function, abbreviated to p.d.f. or simply density function of the r.v. X.

A p.d.f. has the following properties:

(i) $f(x) \geq 0,$ for all x

(ii) $\int_{-\infty}^{\infty} f(x)dx = 1$

(iii) The probability that X takes on a value in the interval $[c, d], c < d$ is given by

$$P(c < x \leq d) = F(d) - F(c)$$

$$= \int_{-\infty}^{d} f(x)dx - \int_{-\infty}^{c} f(x)dx = \int_{c}^{d} f(x)dx$$

JOINT DISTRIBUTIONS

The distribution of two or more random variables which are observed simultaneously when an experiment is performed, is called their joint distribution. It is n customary to call the distribution of a single r.v as univariate. Likewise, a distribution involving two, thee or many r.v. simultaneously is referred to as bivariate, trivariate or multivariate.

Bivariate Distribution Function

Let X and Y be two r.v. defined on the same sample space S. Then the function F(x,y) defined by $F(x, y)$ = P(X $\leq x\ and\ Y \leq y)$, where F(x,y) gives the probability that X will take on a value less than or equal to x and, at the same time, Y will takes on a value less than or equal to y, is called a bivariate or joint distribution function of X and Y.

Bivariate Probability Function

Let X and Y be two discrete r.v defined on the some sample space S, X taking the values $x_1 x_2,\ldots,X_m$ and Y taking the values of $y_1 y_2,\ldots y_n$. Then the probability that X takes on the value x_i and, at the same time Y takes on the value y_j, *denoted by* $f(x_i, y_j)$ Or P_{ij}, *is* defined tpo be the joint probability function, also called bivariate probability function $f(x, y)$ is a function whose value at the points (x_i, y_j) = P (X = x_i *and* $Y = y_j$),

$i=1,2,\ldots\ldots m.$, $j = 1,2,\ldots,n.$

A joint probability function has the following properties:

(i) $f(x_i, y_j) > 0$, for all (x_i, y_j), i.e. for I = 1,2, ...,m; j = 1,2, ..., n.

(ii) $\sum_i \sum_j f(x_i, y_j) = 1$

Marginal Probability Function

From the joint probability function for (X, Y), we can obtain the individual probability function of X and Y. Such individual probability functions are called marginal probability functions

Let $f(x, y)$ be the joint probability function of two discrete r.v. X and Y. Then the marginal probability function of X is defined as

$$G(x_i) = \sum_{j=1}^{n} f(x_i, y_i)$$

$$= f(x_i, y_1) + f(x_i, y_2) + \ldots + f(x_i, y_n)$$

as x_i must occur either with y_1 *or* y_2 *or* $\ldots$ *or* y_n.

$= \mathrm{P}(\mathrm{X} = x_i)$;

Conditional Probability Function

Let X and Y be discrete r. v. with joint probability function $f(x, y)$. Then the Conditional probability function for X given Y = y, denoted as $f(x|y)$ is defined by

$f((x_i|y_j)$ = P (X = $x_i|Y = y_j$)

$= \frac{P(X = x_i \, and \, Y = y_j)}{P(Y = yx_j)}$

$= \frac{f(x_i, y_j}{g(x_i)}$, where g(x) =>0.

Independence

Two discrete r.v. X and Y are said to be statistically independent, if and only if, for all possible pairs of values ($x_i, y_j)$ by the joint probability function $f(x, y)$ can be expressed as the product of the two marginal probability functions. That is, X and Y are independent, if

$f(x, y) = P(X = x_i \, and \, Y = y_j)$

= P(X = x_i) . P (Y = y_j) for all *i* and *j*.

Continuous Bivariate Distribution

The bivariate probability density function of continuous r.v. X and Y is an integrable function $f(x, y)$ satisfying the following properties:

(i) $f(x, y) \geq o \, for \, all (x, y)$

(ii) $\int_{-\infty}^{\infty} \int_{-\infty}^{\infty} f(x, y) \, dx \, dy = 1$, and

(iii) P(a $\leq X \leq b, c \leq Y \leq d) = \int_a^b \int_c^d f(x, y) \, dy \, dx$

The distribution function (d, f) of the bivariate r.v (X, Y) is defined by

$$F(x, y) = \int_{-\infty}^{x}\int_{-\infty}^{y} f(u,v)\,dv\,du.$$

The marginal p.d.f. of the continuous r.v. X is

$$g(x) = \int_{-\infty}^{\infty} f(x,y)\,dy$$

And that of the r.v. Y is

$$h(y) = \int_{-\infty}^{\infty} f(x,y)\,dx$$

That is the marginal p.d.f. of any of the variable is obtained by integrating out the other variable from the joint p.d.f. between the limits $-\infty$ *and* $+\infty$.

The conditional p.d.f. of the continuous r.v. X given that Y takes the value y, is defined to be $f(x|y) = \frac{f(x,y)}{h(y)}$,

Where $f(x,y)$ and h(y) are respectively the joint p.d.f. of X and Y, and the marginal p.d.f. of Y, and h(y) > 0. Similarly, the conditional p.d.f. of the continuous r.v. Y given that X = x, is

$$f(y|x) = \frac{f(x,y)}{g(x)}$$

, proved that g(x) > 0..

Finally two continuous r.v. X and Y are said to statistically independent, if and only if their joint density $f(x,y)$ can be factored in the form f(x, y) = g(x). h(y) for all possible values of X and Y.

CHAPTER 4

EXPECTATION

INTRODUCTION

Earlier we have discussed about sample mean. Which is the arithmetic mean of the given data. Suppose two coins are tossed 10 times and x is the number of heads that occur per toss, then the values of x are $0, 1 \, and \, 2.$ Suppose that the experiment yields no heads, one head and two heads a total of 3, 3, and 4 times, respectively. Then the average number of heads per toss of the two coins is then

$$\frac{(0)(3)+(1)(3)+(2)(4)}{10}=1.1$$

This is an average value of the data and so far it is not a possible outcome of $\{0,1,2\}.$ Hence, an average is not essentially a probable outcome for the trial. For example, a salesman's average monthly income is not likely to be equal to any of his monthly paychecks.

Let us now reshuffle our working out for the average number of heads so as to have the following corresponding form

$$(0)\left(\frac{3}{10}\right)+(1)\left(\frac{3}{10}\right)+2\left(\frac{4}{10}\right)=1.1$$

The numbers $\frac{3}{10}, \frac{3}{10}, and \, \frac{4}{10}$ are the fractions of the total tosses resulting in $0, 1,$ and 2 heads, respectively. These fractions are also relative frequencies for the different values of X in our trial. Actually, then, we can compute the mean, or average, of a set of data by knowing the different values that occur and their relative frequencies, without any information of the total number of observations in our set of data. Therefore, if $\frac{3}{10}$ of the tosses results in no heads, $\frac{3}{10}$ of the tosses results in one head, and $\frac{4}{10}$ of the tosses results in two heads, the mean number of heads per toss would be 1.1 no matter whether the total numbers of tosses were 10,100,1000, or even 10,000.

This process of relative frequencies is used to estimate the average number of heads per toss of two coins that we might look forward in the long run.

We shall refer to this average value as the mean of the random variable X or the mean of the probability distribution of X and write it as μ_x or simply as μ when it is clear to which random variable we refer. It is also common among statisticians to refer to this mean as the mathematical expectation, or the expected value of the random variable X, and denote it as $E(X)$.

Assuming that 1 fair coin was tossed twice, we find that the sample space for our experiment is $S=\{HH,HT,TH,TT\}$.

Since the 4 sample points are equally likely, it follows that

$$P(X=0)=P(TT)=\frac{1}{4}.$$

$$P(X=1)=P(TH)+P(HT)=\frac{1}{2},$$

and $P(X=2)=P(HH)=\frac{1}{4}$

where elements, say TH, indicate that the first toss resulted in a tail followed by a head on the second toss. Now, these probabilities are just the relative frequencies for the given event in the long run, therefore,

$$\mu = E(x) = (0)\left(\frac{1}{4}\right) + (1)\left(\frac{1}{2}\right) + (2)\left(\frac{1}{4}\right) = 1$$

This result means that a person who tosses 2 coins over and over again will, on the average, get 1 head per toss.

The technique explained above for calculating the expected number of heads per toss of 2 coins suggests that the mean, or expected value, of any discrete random variable may be obtained by multiplying each of the values of $x_1, x_2, \ldots, x_n$ the random variable X by its corresponding probablity $f(x_1), f(x_2), \ldots, f(x_n)$ and summing the products. This is true, however, only if the random variable is discrete. In the case of continuous random variables, the definition of an expected value is basically the same with summations replaced by integrations.

Let x be a simple random variable, that is, a discrete random variable taking on only finitely many possible values. Now we can define Expectation (also called expected value, average value, mean value or simply mean) of x as

$$E(x) = x_1 f(x_1) + x_2 f(x_2) + x_3 f(x_3) + \ldots + x_n f(x_n) + \ldots$$

$$E(x) = \sum_{i=1}^{\infty} x_i f(x_i),$$ provided it converges absolutely.

The sum converges absolutely if and only if $\Sigma|x| f(x)$ is finite.

When X takes on only finite number of values, we have $E(x) = \sum_{i=1}^{n} x_i f(x_i),$ which may be regarded as a weighted mean of the variable's possible values $x_1, x_2, \ldots, x_n,$ each being weighted by the respective probability. In case the values are equally likely, $E(x) = \frac{\Sigma x_i}{n},$ which reprsents the ordinary arithmetic mean of the n possible values, $E(x)$ is also called the mean of X and is usually denoted by the letter μ. It should be noted that $E(x)$ is the average value of the random variable X over a very large number of trials. So we can write it as

$E(x)=\sum xf(x)$ if X is discrete, and (1)

$\mu = E(x) = \int_{-\infty}^{\infty} x f(x)dx$ if x is continuous (2)

Provided the integral converges absolutely, i.e. $\int_{-\infty}^{\infty} |x| f(x)dx$ is finite.

Clearly the definition of mathematical expectation in the case of a continuous random variable is essentially the same with summation being replaced by integral.

Expectation of a Function of a Random Variable

Let $H(x)$ be a function of the random variable X. Then $H(x)$ is also a random variable and also has an expected value, as any function of a r.v. is also a r.v. If X is a discrete r.v. with p.d. $f(x)$ then, since $H(x)$ takes the value $H(x_i)$ when $X = x_i$, the expected value of the function $H(X)$ is

$$E[H(X)] = H(x_1)f(x_1) + H(x_2)f(x_2) + \ldots + H(x_n)f(x_n)$$

$= \sum_i H(x_i)f(x_i),$ provided the series converges absolutely.

Similarly, if X is a continuous r.v, with p.d.f. $f(x),$ then

$E[H(X)] = \int_{-\infty}^{\infty} H(x_i)f(x_i)dx$, provided the integral existed.

In particular, if $H(X) = x^2,$ then $E(X^2) = \sum x_i^2 f(x).$

It is relevant to note that $E(x^2)$ is not the same as $[E(X)]^2$. Now to understand concept better, we will see some examples related to mean.

Example: A lot containing 7 components is sampled by a quality inspector; the lot contains 4 good components and 3 defective components.

A sample of 3 is taken by the inspector. Find the expected value of the number of good components in this sample.

Solution: Let X represent the number of good components in the sample. The probability distribution of X is

$$f(x)=\frac{\binom{4}{x}\binom{3}{3-x}}{\binom{7}{3}}, \qquad x=0,1,2,3.$$

Simple calculation yields

$$f(0)=\frac{1}{35},\ f(1)=\frac{12}{35}, f(2)=\frac{18}{35}\ and\ f(3)=\frac{4}{35}.$$

So

$$E(x)=(0)\left(\frac{1}{35}\right)+(1)\left(\frac{12}{35}\right)+(2)\left(\frac{18}{35}\right)+(3)\left(\frac{4}{35}\right)=\frac{12}{7}=1.7$$

So we can arrive at a result that if a sample size 3 is selected at random over and over again from a lot of 4 good components and 3 defective components, it will contain, on average, 1.7 good components.

Example: A salesperson for a medical device company has two appointments on a given day. At the first appointment, he believes that he has a 70% chance to make the deal, from which he can earn \$1000 commission if successful. On the other hand, he thinks he only has a 40% chance to make the deal at the second appointment, from which, if successful, he can make \$1500. What is his expected commission based on his own probability belief? Assume that the appointment results are independent of each other.

Solution: First, we know that the salesperson, for the two appointments, can have 4 possible commission totals: $\$0, \$1000, \$1500, and\ \$2500.$ We then need to calculate their associated probabilities. By independence, we obtain

$$f(0)=(1-0.7)(1-0.4)=0.8,$$

$$f(2500)=(0.7)(0.4)=0.28,$$

$f(1000) = (0.7)(1-0.4) = 0.42,$

$f(1500) = (1-0.7)(0.4) = 0.12,$

Therefore, the expected commission for the salesperson is

$E(X) = (0)(0.18) + (1000)(0.42) + (1500)(0.12) + (2500)(0.28) = \$1300$

Examples 1 and 2 are considered to permit the reader to achieve some insight into what we mean by the expected value of a random variable. In both cases the random variables are discrete. We follow with an example involving a continuous random variable, where an engineer is interested in the mean life of a certain type of electronic device. This is an illustration of a time to failure problem that occurs often in practice. The expected value of the life of a device is an important parameter for its evaluation.

***Example*:** Let X be the random variable that denotes the life in hours of a certain electronics device. The probability density function is

$$f(x) = \begin{cases} \dfrac{20,000}{x^3}, & x > 100, \\ 0, & elsewhere. \end{cases}$$

Find the expected life of this type of device.

***Solution*:** Using (2), we can write it as

$$\mu = E(X) = \int_{100}^{\infty} x \cdot \frac{20,000}{x^3} dx$$

$$= \int_{100}^{\infty} \frac{20,000}{x^2} dx$$

After simplifying, we obtain

$E(X) = 200.$

Therefore, we can expect this type of device to last, on average 200 hours.

***Example*:** What is the mathematical expectation of the number of heads when 3 fair coins are tossed?

***Solution*:** Let the r.v. X represents the number of heads when three fair coins are tossed. Then X has the following probability distribution.

x	0	1	2	3
$f(x)$	$\frac{1}{8}$	$\frac{3}{8}$	$\frac{3}{8}$	$\frac{1}{8}$

Hence the mathematical expectation of X is

$$E(X)=0\times\frac{1}{8}+1\times\frac{3}{8}+2\times\frac{3}{8}+3\times\frac{1}{8}=1.5$$

It should be noted that $E(X)$ is1.5, which is not an integer and not any value X could actually have. It merely means that, if the experiment of tossing 3 coins is repeated a very large number of times, we would expect on the average to get 1.5 heads.

Example: What is the expectation of the number of failures preceding the first success in an infinite series of independent trials with constant probability of success?

Solution: Let the r.v. X denote the number of failures preceding the first success. Then as X takes the values $0,1,2,3,\ldots,$ the respective probablities are $p,qp,q^2p,q^3p,\cdots$ where $q=1-p.$

Hence $E(X)=x_1 f(x_1)+x_2 f(x_2)+x_3 f(x_3)+\ldots$

$$E(X)=0.p+1.qp+2.q^2p+3.q^3p+\ldots$$

$$=qp\left(1+2q+3q^2+\ldots\right)$$

$$=qp(1-q)^{-2}$$

$$=qp(p)^{-2}=\frac{q}{p}$$

Example: If it rains, an umbrella salesman can earn $\$30$ per day. If it is fair, he can lose $\$6$ per day. What is his expectation if the probability of rain is 0.3?

Solution: Let X represents the number of dollars the salesman earns. Then X is a r.v. with possible values $30\,and\,-6,$ where -6 corresponds to the fact that salesman loses, and the corresponding probabilities are $0.3\, and\, 0.7$ respectively.

Hence $E(X) = 30 \times 0.3 + (-0.6) \times 0.7$

$E(X) = \$4.80\, per\, day.$

Example: A man draws 2 balls from a bag containing 3 white and 5 black balls. If he receives $\$70$ for every white ball he draws and $\$7$ for every black ball, finds his expectation.

Solution: Two balls from a bag containing 3 white and 5 black balls can be drawn in the following three mutually exclusive ways:

2 white balls

1 white and 1 black ball

2 black balls

Let p denote the probability of drawing 2 balls

Then $p_1 = \binom{3}{2} \div \binom{8}{2} = \frac{3}{28}$

$$p_2 = \binom{3}{1}\binom{5}{1} \div \binom{8}{2} = \frac{15}{28}$$

$$p_3 = \binom{5}{2} \div \binom{8}{2} = \frac{10}{28}$$

Let X denote the amount to be received. Then

$x_1 = 2 \times 70 + 0 \times 7 = \$140$

$x_2 = 1 \times 70 + 1 \times 7 = \$77$

$x_3 = 0 \times 70 + 2 \times 7 = 14$

Hence the required expectation $= x_1 p_1 + x_2 p_2 + x_3 p_3$

$$= \frac{3}{28} \times 140 + \frac{15}{28} \times 77 + \frac{10}{28} \times 14$$

$= 15 + 41.25 + 5$

$= \$61.25$

Example: Find the expected value of the r.v. X having the probability density function

$$f(x) = \begin{cases} 2(1-x), & 0 < x < 1 \\ 0, & \textit{elsewhere.} \end{cases}$$

Solution: Since probability density function is given so

$$E(X) = \int_{-\infty}^{\infty} x\, f(x)\, dx$$

$$= 2\int_{0}^{1} x(1-x)\, dx$$

$$= 2\left[\frac{x^2}{2} - \frac{x^3}{3}\right] = 2\left[\frac{1}{2} - \frac{1}{3}\right] = \frac{1}{3}$$

Now let us consider a new random variable $g(X)$, which depends on X; that is, each value of $g(X)$ is determined by the value of X. For instance, $g(X)$ might be X^2 or $3X-1$, and whenever X assumes the value 2, $g(X)$ assumes the value $g(2)$. In particular, *if* X is a discrete random variable with probability distribution $f(x)$, for $x = -1, 0, 1, 2,$ and $g(X) = X^2$, then

$$P\left[g(X)=0\right] = P(X=0) = f(0),$$

$$P\left[g(X)=1\right] = P(X=-1) + P(X=1) = f(-1) + f(1),$$

$$P\left[g(X)=4\right] = P(X=2) = f(2),$$

And so the probability distribution of $g(X)$ may be written as

$g(x)$	$P[g(X)=g(x)$
0	$f(0)$
1	$f(-1)+f(1)$
4	$f(2)$

By the definition of the expected value of a random variable, we obtain

$$\mu_{g(x)}=E\left[g(x)\right]=0f(0)+1\left[f(-1)+f(1)\right]+4f(2)$$

$$=(-1)^2 f(-1)+(0)^2 f(0)+(1)^2 f(1)+(2)^2 f(2)=\sum_x g(x)f(x).$$

So we can generalize the above theorem written in (1) and (2).

Let X be a random variable with probability distribution $f(x)$. The expected value of the random variable $g(X)$ is

$$\mu_{g(x)}=E\left[g(X)\right]=\sum_x g(x)f(x).$$

If X is discrete, and

$$\mu_{g(x)}=E\left[g(X)\right]=\int_{-\infty}^{\infty} g(x)f(x)dx.$$

If X is continuous.

***Example*:** Suppose that the number of cars X that pass through a car wash between 4:00 P.M. and 5:00 P.M. on any sunny Friday has the following probability distribution:

x	4	5	6	7	8	9
$P(X=x)$	$\frac{1}{12}$	$\frac{1}{12}$	$\frac{1}{4}$	$\frac{1}{4}$	$\frac{1}{6}$	$\frac{1}{6}$

Let $g(X)=2X-1$ represent the amount of money, in dollars, paid to the attendant by the manager. Find the attendant's expected earnings for this particular time period.

Solution: We can write it by the generalized theorem

$$E[g(X)]=E(2X-1)=\sum_{x=4}^{9}(2x-1)f(x)$$

$$=(7)\left(\frac{1}{12}\right)+(9)\left(\frac{1}{12}\right)+11\left(\frac{1}{4}\right)+13\left(\frac{1}{4}\right)+15\left(\frac{1}{6}\right)+17\left(\frac{1}{6}\right)=12.67$$

Example: Let X be a random variable with density function

$$f(x)=\begin{cases}\frac{x^2}{3}, & -1<x<2,\\ 0, & elsewhere.\end{cases}$$

Find the expected value of $g(X)=4X+3$.

Solution: Since it is a continuous function as probability density function is provided, we can write it by generalized theorem, by simply multiplying the function with $4x+3$.

$$E(4X+3)=\int_{-1}^{2}(4x+3).\frac{x^2}{3}dx$$

$$=\frac{1}{3}\int_{-1}^{2}(4x^3+3x^2)dx$$

After simplifying result, we will obtain

$$E(4X+3)=8$$

We shall now extend our concept of mathematical expectation to the case of two random variables X and Y with joint probability distribution $f(x,y)$.

Let X and Y be random variables with joint probability distribution $f(x,y)$. The mean, or expected value, of the random variable $g(X,Y)$ is

$$\mu_g(X,Y)=E\left[g(X,Y)\right]=\sum_x\sum_y g(x,y)f(x,y)$$ If $X\ and\ Y$ are discrete,

and

$$\mu_g(X,Y)=E\left[g(X,Y)\right]=\int_{-\infty}^{\infty}\int_{-\infty}^{\infty}g(x,y)f(x,y)\,dx\,dy$$

if X and Y are continuous.

Example: Let $X\ and\ Y$ be the random variable with joint probability distribution given table below. Find expected value of $g(X,Y)=XY$.

$f(x,y)$		X			Row Total
0		1	2		
y	0	$\frac{3}{28}$	$\frac{9}{28}$	$\frac{3}{28}$	$\frac{15}{28}$
	1	$\frac{3}{14}$	$\frac{3}{14}$	0	$\frac{3}{7}$
	2	$\frac{1}{28}$	0	0	$\frac{1}{28}$
Column Total		$\frac{5}{14}$	$\frac{15}{28}$	$\frac{3}{28}$	1

Solution: From above definition, we can write

$$E(XY)=\sum_{x=0}^{2}\sum_{y=0}^{2}xy\,f(x,y)$$

$$=(0)(0)f(0,0)+(0)(1)f(0,1)+(1)(0)f(1,0)$$

$$+(1)(1)f(1,1)+(2)(0)f(2,0)$$

$$=f(1,1)=\frac{3}{14}$$

Example: Find $E\left(\frac{Y}{X}\right)$ for the given density function

$$f(x,y)=\begin{cases}\frac{x\left(1+3y^2\right)}{4}, & 0<x<2, 0<y<1,\\ 0, & elsewhere\end{cases}$$

Solution: We have

$$E\left(\frac{Y}{X}\right)=\int_0^1\int_0^2\frac{y\left(1+3y^2\right)}{4}dx\,dy$$

$$=\int_0^1\frac{y+3y^3}{2}dy$$

$$=\frac{5}{8}$$

Variance of Random Variable

The mean, or expected value, of a random variable X is of special importance in statistics because it describes where the probability distribution is centered. By itself, however, the mean does not give an adequate description of the shape of the distribution. We also need to characterize the variability in the distribution.

Now if $H(X)=(X-\mu)^2$, where μ is the population mean, then $E(X-\mu)^2=\Sigma(x_i-\mu)^2 f(x)$.

We call this expected value of the *variance* and denoted it by $Var(X)$ or σ^2.

That is $\sigma^2=E(X-\mu)^2=E\left(X^2\right)-\left[E(X)\right]^2$.

The positive square root of the variance, as before, is called the standard deviation. It is useful to note the following important results about variance.

$Var(X)$ cannot be negative.

$Var(a)=0,$ where a is a constant.

$Var(aX)=a^2\,Var(X)$

$Var(aX+b)=a^2Var(X)$, where a and b are the contants.

Example: Let X have the following probability distribution:

x	**1**	**2**	**3**	**4**	**5**
$f(x)$	0.2	0.3	0.2	0.2	0.1

Find the probability functions of $3X-1, X^2\ and\ X^2+2;$ and find $E(3X-1), E(X^2)$ and $E(X^2+2)$

Solution:

x	**1**	**2**	**3**	**4**	**5**
$f(x)$	0.2	0.3	0.2	0.2	0.1
$3X-1$	**2**	**5**	**8**	**11**	**14**

$E(3X-1)=\Sigma(3x_i-1)f(x_i)$

$=2\times0.2+5\times0.3+8\times0.2+11\times0.2+14\times0.1$.

$=0.4+1.5+1.6+2.2+1.4=7.1$

Similarly the probability distribution for $H(X)=X^2$ is

x	**1**	**2**	**3**	**4**	**5**
$f(x)$	0.2	0.3	0.2	0.2	0.1
X^2	**1**	**4**	**9**	**16**	**25**

$E(X^2)=\Sigma x_i^2 f(x_i)$

$= 1\times 0.2 + 4\times 0.3 + 9\times 0.2 + 16\times 0.2 + 25\times 0.1$

$= 0.2 + 1.2 + 1.8 + 3.2 + 2.5 = 8.9$

Similarly for $E\left(X^2+2\right) = \Sigma(x_i^2+2)\,f\left(x_i\right)$

$= 3\times 0.2 + 6\times 0.3 + 11\times 0.2 + 18\times 0.2 + 27\times 0.1$

$= 0.6 + 1.8 + 2.2 + 3.6 + 2.7$

$= 10.9$

***Example*:** Let X be a r.v. with probability density function

$$f(x) = \begin{cases} \frac{3}{4}(3-x)(x-5), & 3 < x < 5 \\ 0, & elsewhere \end{cases}$$

Calculate the arithmetic mean, variance and standard deviation of X.

***Solution*:** First we will calculate $E(X)$ and then $E\left(X^2\right)$.

$$E(X) = \int_{-\infty}^{\infty} x\,f(x)\,dx$$

$$= \frac{3}{4}\int_{3}^{5} x.(3-x)(x-5)\,dx$$

$$= \frac{3}{4}\int_{3}^{5}\left(-x^3+8x^2-15x\right)dx$$

$$= \frac{3}{4}\left[-\frac{x^4}{4}+\frac{8x^3}{3}-\frac{15x^2}{2}\right]_3^5$$

After simplifying, we will obtain

$$= \frac{3}{4}\left(\frac{64}{12}\right) = 4$$

Again now for

$$E\left(X^2\right) = \int_{-\infty}^{\infty} x^2\,f(x)\,dx$$

$$=\frac{3}{4}\int_{3}^{5}x^2(3-x)(x-5)\,dx$$

$$=\frac{3}{4}\int_{3}^{5}(-x^4+8x^3-15x^2)\,dx$$

$$=\frac{3}{4}\left[-\frac{x^5}{5}+\frac{8x^4}{4}-\frac{15x^3}{3}\right]_3^5$$

After simplifying, we will obtain

$$=\frac{3}{4}\left(\frac{108}{5}\right)=\frac{81}{5}$$

Now applying formula for variance

$$Var(X)=E(X^2)-[E(X)]^2$$

$$=\frac{81}{5}-(4)^2$$

$$=\frac{1}{5}=0.2$$

So $S.D(X)=\sqrt{0.2}=0.447$

***Example*:** The continuous r.v has probability density function $f(x)$. where

$$f(x)=\frac{3}{4}(1+x^2)\,for\,0<x<1.$$

Now if $E(X)=\mu$ and $Var(X)=\sigma^2$, find $P(|X-\mu|<\sigma)$.

***Solution*:** $E(X)=\int_{-\infty}^{\infty}x\,f(x)\,dx$

$$=\frac{3}{4}\int_{0}^{1}x\,(1+x^2)\,dx$$

$$= \frac{3}{4}\int_0^1 (x + x^3)\,dx$$

$$= \frac{3}{4}\left[\frac{x^2}{2} + \frac{x^4}{4}\right]_0^1$$

$$= \frac{3}{4}\left[\frac{1}{2} + \frac{1}{4}\right] = \frac{9}{16} = 0.5625$$

And

$$E\left(X^2\right) = \int_{-\infty}^{\infty} x^2\, f\left(x\right)dx$$

$$= \frac{3}{4}\int_0^1 x^2\,(1 + x^2)\,dx$$

$$= \frac{3}{4}\int_0^1 (x^2 + x^4)\,dx$$

$$= \frac{3}{4}\left[\frac{x^3}{3} + \frac{x^5}{5}\right]_0^1$$

$$= \frac{3}{4}\left[\frac{1}{3} + \frac{1}{5}\right] = \frac{2}{5} = 0.4$$

$$Var\left(X\right) = E\left(X^2\right) - \left[E\left(X\right)\right]^2$$

$$= \frac{2}{5} - \left(\frac{9}{16}\right)^2 = \frac{107}{1280} = 0.0836$$

$$S.D\left(X\right) = \sqrt{0.0836} = 0.289$$

$$P\left(\left|X - \mu\right| < \sigma\right) = P\left(\,-\sigma < X - \mu < \sigma\right)$$

$$= P\;(\mu - \sigma < X < \mu + \sigma)$$

$$= P(0.5625 - 0.289 < X < 0.5625 + 0.289)$$

$= P(0.2735 < X < 0.8515)$, and

$$P(0.2735 < X < 0.8515) = \frac{3}{4}\int_{0.2735}^{0.8515}(1+x^2)dx$$

$$= \frac{3}{4}\left[x + \frac{x^3}{3}\right]_{0.2735}^{0.8515}$$

After simplifying, we will obtain

$P(0.2735 < X < 0.8515) = 0.8527$

Hence $P(|X-\mu| < \sigma) = 0.8527$

Example: A continuous r.v. X has the probability denisty function

$$f(x) = \frac{3}{4}x(2-x), \quad 0 \le x \le 2.$$

$$= 0, \qquad elsewhere$$

Find the first four moments about the mean and the coefficient of skewness.

Solution: We first calculate the moments about the origin as:

$$\mu_1' = E(X) = \int_{-\infty}^{\infty} x f(x)dx$$

$$= \frac{3}{4}\int_0^2 x(2x - x^2)dx$$

$$= \frac{3}{4}\left[\frac{2x^3}{3} - \frac{x^4}{4}\right]_0^2$$

$$= \frac{3}{4}\left[\frac{16}{3} - \frac{16}{4}\right]$$

$$= \frac{3}{4}\left(\frac{16}{12}\right) = 1$$

$$\mu_2' = E\left(X^2\right) = \int_{-\infty}^{\infty} x^2 f\left(x\right)dx$$

$$= \frac{3}{4}\int_0^2 x^2\left(2x - x^2\right)dx$$

$$= \frac{3}{4}\left[\frac{2x^4}{4} - \frac{x^5}{5}\right]_0^2$$

$$= \frac{3}{4}\left[8 - \frac{32}{5}\right]$$

$$= \frac{3}{4}\left(\frac{8}{5}\right) = \frac{6}{5}$$

$$\mu_3' = E\left(X^3\right) = \int_{-\infty}^{\infty} x^3 f\left(x\right)dx$$

$$= \frac{3}{4}\int_0^2 x^3\left(2x - x^2\right)dx$$

$$= \frac{3}{4}\left[\frac{2x^5}{5} - \frac{x^6}{6}\right]_0^2$$

$$= \frac{3}{4}\left[\frac{64}{5} - \frac{64}{6}\right]$$

$$= \frac{3}{4}\left(\frac{64}{30}\right) = \frac{8}{5}$$

$$\mu_4' = E\left(X^4\right) = \int_{-\infty}^{\infty} x^4 f\left(x\right)dx$$

$$= \frac{3}{4}\int_0^2 x^4\left(2x - x^2\right)dx$$

$$=\frac{3}{4}\left[\frac{2x^6}{6}-\frac{x^7}{7}\right]_0^2$$

$$=\frac{3}{4}\left[\frac{64}{3}-\frac{128}{7}\right]$$

$$=\frac{3}{4}\left(\frac{64}{21}\right)=\frac{16}{7}$$

Then, we find the moments about the mean as

$$\mu_1=0$$

$$\mu_2=\mu_2^{'}-\left(\mu_1^{'}\right)^2=\frac{6}{5}-(1)^2=\frac{1}{5}$$

$$\mu_3=\mu_3^{'}-3\,\mu_1^{'}\,\mu_2^{'}+2\left(\mu_1^{'}\right)^3$$

$$=\frac{8}{5}-3(1)\left(\frac{6}{5}\right)+2(1)^3=\frac{8}{5}-\frac{18}{5}+2=0;$$

$$\mu_4=\mu_4^{'}-4\,\mu_1^{'}\,\mu_3^{'}+6\left(\mu_1^{'}\right)^2\mu_2^{'}-3\left(\mu_1^{'}\right)^4$$

$$=\frac{16}{7}-4(1)\left(\frac{8}{5}\right)+6(1)^2\left(\frac{6}{5}\right)-3(1)^4=\frac{3}{35}$$

Now, the coefficient of skewness is

$$\gamma_1=\frac{E\left(X-\mu\right)^3}{\sigma^3}=0$$

Properties of Expected Values

The important properties of the expected values of a r.v. are given below:

1. If a is a constant, then $E\left(a\right)=a$

Proof: Let X be a discrete r.v. with probability distribution $\left[x_i, f\left(x_i\right)\right], i=1,2,\ldots,n.$

Then $E(X)$ for $X = a$ is given by

$$E(a) = \sum_{i=1}^{n} a\, f(x_i)$$

$$= a\, f(x_1) + a_2\, f(x_2) + \ldots + a\, f(x_n)$$

$$= a\left[f(x_1) + f(x_2) + \ldots f(x_n) \right]$$

$$= a \sum_i f(x_i)$$

$$= a \qquad \therefore \sum_i f(x_i) = 1$$

2. If X is a discrete r.v and if a and b are constants, then $E(aX + b) = a\,E(X) + b.$

Proof: Let the probability distribution of the r.v. X be $\left[x_i, f(x_i) \right], i = 1, 2, \ldots, n.$

Then by definition of expected value, we have

$$E(aX + b) = \sum_{i=1}^{n} (ax_i + b) f(x_i)$$

$$= (a\,x_1 + b) f(x_1) + (a\,x_2 + b) f(x_2) + \ldots (ax_n + b) f(x_n)$$

$$= a\left[x_1 f(x_1) + x_2\, f(x_2) + \ldots x_n f(x_n) \right] + b\left[f(x_1) + f(x_2) + \ldots f(x_n) \right]$$

$$= a \sum_i x\, f(x_i) + b \sum_i f(x_i)$$

$$= a\,E(X) + b$$

This result is often called the expected value of linear transformation of the r.v. X.

If $b = 0,$ then $E(aX) = a\,E(X).$

If $a = 1,\ b = -\mu,$ then $E(X - \mu) = E(X) - \mu = \mu - \mu = 0$

That is, the expected value of the deviation of any r.v. from its mean is always equal to zero.

3. The expected value of the sum of any two random variables is equal to the sum of their expected values, i.e. $E(X+Y)=E(X)+E(Y)$

Proof: Let X and Y be two discrete r.v.'s defined on the same sample space S. Let X assume m values $x_1, x_2, \ldots, x_m$ with probabilities $g(x_1), g(x_2), \ldots, g(x_m)$ and Y assume n values $y_1, y_2, \ldots, y_n$ with probabilties $h(y_1), h(y_2), \ldots, h(y_n)$. The sum $X+Y$ is a r.v. taking the values $x_i + y_j$ with probabilities $f(x_i, y_j)$ for all combinations of values of i and j. Hence by definition, we have

$$E(X+Y)=\sum_i\sum_j (x_i + y_j) f(x_i, y_j)$$

$$=\sum_i\sum_j (x_i) f(x_i, y_j) + \sum_i\sum_j (y_j) f(x_i, y_j)$$

But

$$\sum_i\sum_j (x_i) f(x_i, y_j) = \sum_i x_i \sum_j f(x_i, y_j)$$

$$=\sum_i x_i \left[f(x_i, y_1) + f(x_i, y_2) + \ldots + f(x_i, y_n) \right]$$

$$=\sum_i x_i g(x_i) \quad (\therefore \text{all possible values of y are included in} \sum_j)$$

$$= E(X)$$

Similarly,

$$\sum_i\sum_j (x_i) f(x_i, y_j) = \sum_j y_j \sum_i f(x_i, y_j)$$

$$=\sum_j y_j \left[f(x_1, y_j) + f(x_2, y_j) + \ldots + f(x_m, y_i) \right]$$

$$\sum_j y_j h(y_j) = E(Y)$$

Hence $E(X+Y)=E(X)+E(Y)$

It is interesting to note that this result holds in general, that is, the expected value of any finite number of r.v. is the sum of their expected values. In other words,

$$E(X_1+X_2+X_3+\ldots+X_n)=E(X_1)+E(X_2)+E(X_3)+\ldots E(X_n)$$

or

$$E(\Sigma X_i)=\sum_i[E\ (X_i)]$$

The result also holds for the difference of r.v., i.e.

$$E(X-Y)=E(X)-E(Y).$$

4. The expected value of the product of two independent r.v. is equal to the product of their expected values, i.e. $E(XY)=E(X)E(Y)$.

Proof: Let the r.v. X assume m values $x_1,x_2,\ldots,x_m$ and the r.v. Y, the n values $y_1,y_2,\ldots,y_n$.

Let the probability of Y assuming the value y_j be $h(y_j)$. The product XY is a r.v. taking the value x_iy_j with probabilities $f(x_i,y_j)$. As $X\ and\ Y$ are independent, their joint p.d. $f(x_i,y_j)$ can be factored as $f(x_i,y_j)=g(x_i)h(y_j)$.

Hence $$E(XY)=\sum_i\sum_j x_iy_jf(x_i,y_j)$$

$$=\sum_i\sum_j x_iy_jg(x_i)h(y_j)$$

$$=\sum_i x_ig(x_i)\sum_j y_j\,h(y_j)$$

$$=E(X)E(Y)$$

This result can easily be extended to several independent r.v.

It should be noted that these properties are valid for continuous r.v. in which case the summations are replaced by integrals.

Example: Let be two discrete r.v. with the following joint probability distribution

x / y	2	4
1	0.10	0.15
3	0.20	0.30
5	0.10	0.15

Find $E(X), E(Y), E(X+Y), E(2X-3Y) \, and \, E(XY)$

Solution: Now to determine the expected values of and , we first find the marginal probability distribution and by adding over the columns and rows of the two-way table as below:

x / **y**	**2**	**4**	
1	0.10	0.15	0.25
3	0.20	0.30	0.5
5	0.10	0.15	0.25
	0.40	**0.60**	**1.00**

$$E(X) = \Sigma x_i g(x_i) = 2 \times 0.40 + 4 \times 0.60 = 0.80 + 2.40 = 3.2$$

$$E(Y) = \Sigma y_i h(y_i) = 1 \times 0.25 + 3 \times 0.5 + 5 \times 0.25 = 0.25 + 1.5 + 1.25 = 3$$

$$E(X+Y) = E(X) + E(y)$$

$$= 3.2 + 3 = 6.2$$

$$E(2X - 3Y) = 2E(X) - 3E(Y) = 2(3.2) - 3(3) = -2.6$$

Since $X \, and \, Y$ are independent, therefore

$$E(XY) = E(X).E(Y) = (3.2)(3.0) = 9.6$$

Example: $X \text{ and } Y$ are two independent r.v. such that $g(x) = \frac{1}{3} \text{ for } x = 1,2,3 \text{ and } h(y) = \frac{1}{2} \text{ for } y = 0,1$ If $Z = 2X - Y$, then verify that $E(Z) = 2E(X) - E(y)$.

Solution: Now to determine the expected values of X and Y, we first find the marginal probability distribution $g(x)$ and $h(y)$ by adding over the columns and rows of the two-way table as below:

x y	1	2	3	$h(y)$
0	$\frac{1}{6}$	$\frac{1}{6}$	$\frac{1}{6}$	$\frac{1}{2}$
1	$\frac{1}{6}$	$\frac{1}{6}$	$\frac{1}{6}$	$\frac{1}{2}$
$g(x)$	$\frac{1}{3}$	$\frac{1}{3}$	$\frac{1}{3}$	1.00

$$E(X) = \Sigma x_i g(x_i) = \left(1 \times \frac{1}{3}\right) + \left(2 \times \frac{1}{3}\right) + \left(3 \times \frac{1}{3}\right) = 2$$

$$E(Y) = \Sigma y_i h(y_i) = \left(0 \times \frac{1}{2}\right) + \left(1 \times \frac{1}{2}\right) = \frac{1}{2}$$

$$E(Z) = E(2X - Y) = 2E(X) - E(Y) = 2(2) - \frac{1}{2} = \frac{7}{2}$$

For verification, we find the value of $E(2X - Y)$ directly as below:

$$E[2X - Y] = \sum_i \sum_j (2x_i - y_j) f(x_i, y_j)$$

$$= (2 \times 1 - 0) \times \frac{1}{6} + (2 \times 1 - 1) \times \frac{1}{6} + (2 \times 2 - 0) \times \frac{1}{6}$$

$$+(2\times2-1)\times\frac{1}{6}+(2\times3-0)\times\frac{1}{6}+(2\times3-1)\times\frac{1}{6}$$

$$=\frac{2}{6}+\frac{1}{6}+\frac{4}{6}+\frac{3}{6}+\frac{6}{6}+\frac{3}{6}$$

$$=\frac{21}{6}=\frac{7}{2}$$

Example: Three dice, each numbered in the usual way from one to six, are colored white, red and blue respectively. After casting them. A boy scores in the following way. To the white number he adds twice the red numbers and then subtract the blue number. Thus a white three, a red four and a blue two would score $3+8-2=9$. Assume that the boy casts the dice a large number of times, calculate the mean and variance of the scores.

Solution: Let $X_w, X_r \; and \; X_b$ denote the numbers when the white, red and blue dice are cast respectively; and let Y represents the score a boy gets by casting the dice. Then his score according to the condition given is $Y = X_w + 2X_r - X_b$, and we need $E(Y) and \, Var(Y)$.

Now $E(Y) = E[X_w + 2X_r - X_b]$

$$= E(X_w) + 2E(X_r) - E(X_b)$$

$$=1\times\frac{1}{6}+2\times\frac{1}{6}+3\times\frac{1}{6}+4\times\frac{1}{6}+5\times\frac{1}{6}+6\times\frac{1}{6}=\frac{21}{6}$$

Therefore $E(Y)=\frac{21}{6}+2\left(\frac{21}{6}\right)-\frac{21}{6}=\frac{21}{3}=7$

Again by definition

$$E(Y^2)=E(X_w+2X_r-X_b)^2$$

$$=E(X_w^2)+4E(X_r^2)+E(X_b^2)+4E(X_w X_r)-2E(X_w X_b)-4E(X_r X_b)$$

$$=1\times\frac{1}{6}+4\times\frac{1}{6}+9\times\frac{1}{6}+16\times\frac{1}{6}+25\times\frac{1}{6}+36\times\frac{1}{6}=\frac{91}{6}$$

$$E\left(Y^2\right)=\frac{91}{6}+4\left(\frac{91}{6}\right)+\frac{91}{6}+4\left(\frac{21}{6}\right)\left(\frac{21}{6}\right)-2\left(\frac{21}{6}\right)\left(\frac{21}{6}\right)-4\left(\frac{21}{6}\right)\left(\frac{21}{6}\right)$$

$$=\frac{91}{6}+\frac{364}{6}+\frac{91}{6}+49-\frac{49}{2}-49=\frac{133}{2}$$

Hence $Var\left(Y\right)=\frac{133}{2}-\left(7\right)^2=17.5$

Example: Let X and Y be independent r.v's with joint probability density function

$$f\left(x,y\right)=\frac{x\left(1+3y^2\right)}{4},\ 0<x<2, 0<y<1$$

$$=0,\qquad elsewhere$$

Find $E\left(X\right), E\left(Y\right), E\left(X+Y\right) and\ E\left(XY\right).$

Solution: To determine $E\left(X\right)$ and $E\left(Y\right)$, we first find the marginal $p.d.f$ $g\left(x\right)$ and $h\left(y\right)$ as below:

$$g\left(x\right)=\int_{-\infty}^{\infty}f\left(x,y\right)dy=\int_{0}^{1}\frac{x\left(1+3y^2\right)}{4}dy$$

$$=\frac{1}{4}\left[xy+xy^3\right]_0^1=\frac{x}{2}$$

$$h\left(y\right)=\int_{-\infty}^{\infty}f\left(x,y\right)dx$$

$$=\int_{0}^{2}\frac{x\left(1+3y^2\right)}{4}dx$$

$$=\frac{1}{4}\left[\frac{x^2}{2}+3xy^2\right]_0^2$$

$$=\frac{1}{2}\left(1+3y^2\right)$$

Hence

$$E(X)=\int_{-\infty}^{\infty}x.g(x)\,dx$$

$$=\int_0^2 x\left(\frac{x}{2}\right)dx$$

$$=\frac{1}{2}\left[\frac{x^3}{3}\right]_0^2=\frac{4}{3}$$

And

$$E(Y)=\int_{-\infty}^{\infty}y.h(y)\,dy$$

$$=\frac{1}{2}\int_0^1 y\,(1+3y^2)\,dy$$

$$=\frac{1}{2}\left[\frac{y^2}{2}+\frac{3y^4}{4}\right]_0^1=\frac{1}{2}\left[\frac{1}{2}+\frac{3}{4}\right]=\frac{5}{8}$$

$$E(X+Y)=\int_{-\infty}^{\infty}\int_{-\infty}^{\infty}(x+y)\,f(x,y)\,dx\,dy$$

$$=\int_0^2\int_0^1(x+y)\frac{x\left(1+3y^2\right)}{4}\,dy\,dx$$

$$=\int_0^2\int_0^1\frac{x^2+3x^2y^2}{4}\,dy\,dx+\int_0^2\int_0^1\frac{xy+3xy^3}{4}\,dy\,dx$$

$$=\int_0^2\frac{1}{4}\left[x^2y+x^2y^3\right]_0^1dx+\int_0^2\frac{1}{4}\left[\frac{xy^2}{2}+\frac{3xy^4}{4}\right]_0^1dx$$

$$= \int_0^2 \frac{1}{4}\left(2x^2\right)dx + \int_0^2 \frac{1}{4}\left(\frac{x}{2}+\frac{3x}{4}\right)dx$$

$$= \frac{1}{2}\left[\frac{x^3}{3}\right]_0^2 + \frac{1}{4}\left[\frac{x^2}{4}+\frac{3x^2}{8}\right]_0^2$$

$$= \frac{4}{3}+\frac{5}{8}=\frac{47}{24}$$

$$E(XY) = \int_{-\infty}^{\infty}\int_{-\infty}^{\infty} xy\, f(x,y)\,dx\,dy$$

$$= \int_0^2\int_0^1 xy\frac{x\left(1+3y^2\right)}{4}dy\,dx$$

$$= \int_0^2\int_0^1 \frac{x^2y+3x^2y^3}{4}dy\,dx$$

$$= \int_0^2 \frac{1}{4}\left(\frac{x^2y^2}{2}+\frac{3x^2y^4}{4}\right)_0^1 dx$$

$$= \int_0^2 \frac{1}{4}\left(\frac{5x^2}{4}\right)dx$$

$$= \frac{1}{4}\left[\frac{5x^3}{12}\right]_0^2 = \frac{5}{6}$$

It should be noted that

$$E(X)+E(Y)=\frac{4}{3}+\frac{5}{8}=\frac{47}{24}=E(X+Y)$$

$$E(X)E(Y)=\left(\frac{4}{3}\right)\left(\frac{5}{8}\right)=\frac{5}{6}=E(XY)$$

Covariance of Two Random Variables

The covariance of two r.v. $X \, and \, Y$ is a numerical measure of the extent to which their values tend to increase or decrease together. It is denoted by σ_{xy} or $Cov(X,Y)$, and is defined as the expected value of the product $[X-E(X)][Y-E(Y)]$. That is

$$Cov(X,Y) = E\{[X-E(X)][Y-E(Y)]\}$$

$$= E[XY - X\,E(Y) - Y\,E(X) + E(X)E(Y)]$$

$$= E(XY) - E(X)E(Y) - E(Y)E(X) + E(X)E(Y)$$

$$= E(XY) - E(X)E(Y).$$

If $X \, and \, Y$ are independent, then $E(XY) = E(X)E(Y)$, and $Cov(X,Y) = E(XY) - E(X)E(Y) = 0$

It is very important to note that covariance is zero when the r.v. X and Y are independent but its converse is not generally true. The covariance of a r.v. with itself is obviously its variance.

Variance of the Sum or Difference of Two Random Variables

Let X and Y be two discrete r.v.'s. Then the variance of the r.v. $X+Y$ is defined as

$$Var(X+Y) = Var(X) + Var(Y)$$

$$Var(X-Y) = Var(X) + Var(Y)$$

if X and Y are independent.

Correlation Coefficient of Random Variables

Let X and Y be two r.v. with non-zero variance $\sigma_X^2 \, and \, \sigma_Y^2$. Then the correlation coefficient which is a measure of linear relationship between $X \, and \, Y$ denoted by ρ_{XY} or $Corr(X,Y)$ is defined as

$$\rho_{XY} = \frac{Cov(X,Y)}{\sqrt{Var(X)Var(Y)}}$$

If $X\ and\ Y$ are independent r.v., then ρ_{XY} will be zero, but zero correlation does not necessarily imply independence. The correlation coefficient has the following properties:

- Correlation coefficient is unitless and symmetric in $X\ and\ Y, i.e., \rho_{XY} = \rho_{YX}$
- Correlation coefficient remains unchanged if constants are added to the r.v. or if the r.v. are multiplied by constants having the same sign.
- Correlation coefficient is a number between $-1\ and\ +1$ inclusive. Or we can write it

$$-1 \le \rho \le 1$$

Example: From the following p.d of $X\ and\ Y$, find $Var(X), Var(Y), Cov(X,Y)\ and\ \rho.$

Y X	0	1	2	3	$g(x)$
0	0.05	0.05	0.10	0	0.20
1	0.05	0.10	0.25	0.10	0.50
2	0	0.15	0.10	0.05	0.30
$h(y)$	0.10	0.30	0.45	0.15	1.00

Now $E(X) = \Sigma x_i g(x_i) = 0\times0.20 + 1\times0.50 + 2\times0.30 = 1.10$

Now $E(Y) = \Sigma y_j h(y_j) = 0\times0.10 + 1\times0.30 + 2\times0.45 + 3\times0.15 = 1.65$

$$E(X^2) = \Sigma x_i^2 g(x_i) = 0\times0.20 + 1\times0.50 + 4\times0.30 = 1.70$$

$$E(Y^2) = \Sigma y_j^2 h(y_j) = 0\times0.10 + 1\times0.30 + 4\times0.45 + 9\times0.15 = 3.45$$

Thus, $Var(X) = E(X^2) - [E(X)]^2 = 1.70 - (1.10)^2 = 0.49,$ and

$$Var(Y) = E(Y^2) - [E(Y)]^2 = 3.45 - (1.65)^2 = 0.7275$$

$$E(XY) = \sum_i \sum_j (x_i, y_j) f(x_i, y_j)$$

$$= 1 \times 0.10 + 2 \times 0.15 + 2 \times 0.25 + 4 \times 0.10 + 3 \times 0.10 + 6 \times 0.05$$

$$= 0.10 + 0.30 + 0.50 + 0.40 + 0.30 + 0.30 = 1.90$$

$$Cov(X,Y) = E(XY) - E(X)E(Y) = 1.90 - 1.10 \times 1.65 = 0.085$$

$$\rho = \frac{Cov(X,Y)}{\sqrt{Var(X)Var(Y)}} = \frac{0.085}{\sqrt{(0.49)(0.7275)}} = \frac{0.085}{0.595} = 0.14$$

Example: If $f(x,y) = x^2 + \frac{xy}{3}, \ 0 \le x \le 1, 0 \le y \le 2$

$= 0, \qquad elsewhere,$

Find $Var(X), Var(Y)\, and\, Corr(X,Y).$

Solution: The marginal probability distribution functions are

$$g(x) = \int_0^2 \left(x^2 + \frac{xy}{3} \right) dy = 2x^2 + \frac{2}{3}x,$$

$$and\, h(y) = \int_0^1 \left(x^2 + \frac{xy}{3} \right) dx = \frac{1}{3} + \frac{y}{6}$$

$$E(X) = \int_{-\infty}^{\infty} x g(x)\, dx$$

$$= \int_0^1 x \left(2x^2 + \frac{2x}{3} \right) dx = \frac{13}{18}$$

$$E(Y) = \int_{-\infty}^{\infty} y h(y)\, dy$$

$$= \int_0^2 y\left(\frac{1}{3}+\frac{y}{6}\right)dy = \frac{10}{9}$$

Thus, $Var(X) = E\left[X - E(X)\right]^2 = \int_{-\infty}^{\infty}(x-\mu_x)^2 g(x)dx$

$$= \int_0^1 \left(x - \frac{13}{18}\right)^2 \left(2x^2 + \frac{2x}{3}\right)dx = \frac{73}{1620}$$

$$Var(Y) = E\left[Y - E(Y)\right]^2$$

$$= \int_{-\infty}^{\infty}(y-u_y)^2 h(y)dy$$

$$= \int_0^2 \left(y - \frac{10}{9}\right)^2 \left(\frac{1}{3}+\frac{y}{6}\right)dy = \frac{26}{81}$$

$$Cov(X,Y) = E\left\{\left[X - E(X)\right]\left[Y - E(Y)\right]\right\}$$

$$= \int_0^1\int_0^2 \left(x - \frac{13}{18}\right)\left(y - \frac{10}{9}\right)\left(x^2 + \frac{xy}{3}\right)dy\,dx$$

$$= \int_0^1 (-\frac{2}{9}x^3 + \frac{25}{81}x^2 - \frac{26}{243}x)dx = -\frac{1}{162}$$

Hence $Corr(X,Y) = \dfrac{Cov(X,Y)}{\sqrt{Var(X)Var(Y)}}$

$$= -\frac{\frac{1}{162}}{\sqrt{\left(\frac{73}{1620}\right)\left(\frac{26}{81}\right)}} = -0.05$$

Chebyshev's Inequality

In previous section we already discussed that the variance of a random variable gave us the variability of the observations about the mean. If a random variable has a small variance or standard deviations, we would expect most of the values to be grouped around the mean. Therefore, the probability that the random variable assumes a value within a certain interval about the mean is greater than for a similar random variable with a large standard deviation. If we think of probability in terms of area, we would expect a continuous distribution with a large value of to indicate a greater variability, and therefore we should expect the area to be more spread out. A distribution with small standard deviation should have most of its area close to

The Russian mathematician P. L. Chebyshev (1821–1894) discovered that the fraction of the area between any two values symmetric about the mean is related to the standard deviation. Since the area under a probability distribution curve or in a probability histogram adds to 1, the area between any two numbers is the probability of the random variable assuming a value between these numbers.

The following theorem, due to Chebyshev, gives a conservative estimate of the probability that a random variable assumes a value within k standard deviations of its mean for any real number k.

If X is a r.v. having mean μ and variance $\sigma^2 > 0$, and k is any positive constant, then the probability that a value of X falls within k standard deviations of the mean is atleast $\left(1-\frac{1}{k^2}\right)$.

That is, $$P(\mu - k\sigma < X < \mu + k\sigma) \geq 1 - \frac{1}{k^2},$$

or equivalently, $$P\left[|x-\mu| \geq k\sigma\right] \leq \frac{1}{k^2}.$$

This inequality is due to the Russian mathematician P.L. Chebyshev and it provides a means of understanding how the variance measures variability about the mean of a r.v.

Chebyshev's theorem holds for any distribution of observations, and for this reason the results are usually weak. The value given by the theorem is a lower bound only. That is, we know that the probability of a random variable falling within two standard deviations of the mean can be *no less* than 3/4, but we never know how much more it might actually be. Only when the probability distribution is known can we determine exact probabilities. For this reason we call the theorem a *distribution-free* result. When specific distributions are assumed, as in future chapters, the results will be less conservative. The use of Chebyshev's theorem is relegated to situations where the form of the distribution is unknown.

Moment Generating Function

The moment generating function (m.g.f.) usually denoted by $M_0(t)$, of a random variable X about the origin if it exists, is defined as the expected value of the r.v. e^{tX}, where t is a real variable lying in a neighborhood of zero. That is

$$M_0(t) = E\left(e^{tX}\right) = \sum_{i=1}^{\infty} e^{tx_i} f(x_i), \quad \textit{if X is discrete,}$$

$$= \int_{-\infty}^{\infty} e^{tx} f(x)\,dx, \quad \textit{if X is continuous.}$$

Since $e^{\theta} = 1 + \theta + \frac{\theta^2}{2!} + \frac{\theta^3}{3!} + \ldots,$ therefore for the discrete case

$$M_0(t) = \Sigma\left[1 + tx + \frac{(tx)^2}{2!} + \frac{(tx)^3}{3!} + \ldots + \frac{(tx)^r}{r!} + \ldots\right] f(x)$$

$$= \Sigma f(x) + t\Sigma x f(x) + \frac{t^2}{2!}\Sigma x^2 f(x) + \frac{t^3}{3!}\Sigma x^3 f(x) + \ldots + \frac{t^r}{r!}\Sigma x^r f(x) + \ldots$$

$$= 1 + tE(X) + \frac{t^2}{2!}E\left(X^2\right) + \frac{t^3}{3!}E\left(X^3\right) + \ldots + \frac{t^r}{r!}E\left(X^r\right) + \ldots$$

$$= 1 + t\mu_1' + \frac{t^2}{2!}\mu_2' + \frac{t^3}{3!}\mu_3' + \ldots + \frac{t^r}{r!}\mu_r' + \ldots$$

$$= \sum_{r=0}^{\infty} \frac{t^r}{r!}\mu_r'.$$

Thus we find that the coefficient of $\frac{t^r}{r!}$ in the expansion of $M_0(t)$ is just $E(X^r)$ or μ_r', the rth moment about zero. We call the function $M_0(t)$ the moment generating function because it generates all of the moments of the r.v. X. It should be noted that a moment generating function would exist only if the sum or integral converges for all values of t in a neighborhood of zero.

It is more convenient to find the moments by differentiating the moment generating function r times with respect to t and putting $t = 0$. The moment generating function of X is

$$M_0(t) = 1 + tE(X) + \frac{t^2}{2!}E(X^2) + \frac{t^3}{3!}E(X^3) + \ldots + \frac{t^r}{r!}E(X^r) + \ldots$$

The m.g.f. about the value *a* is defined as the expected value of $e^{t(X-a)}$ and is denoted as

$$M_a(t) = E\left[e^{t(X-a)}\right] = e^{-at}E^{tX} = e^{-at}M_0(t)$$

Thus the m.g.f. about the value *a* is equal to e^{-at} times the m.g.f. about the origin.

SUMMARY AND DISCUSSION

In probability, the expected value of a random variable, obviously, is the long-run average value of repetitions of the experiment it represents. Roughly, the law of large numbers states that the arithmetic mean of the values nearly converges to the expected value as the number of repetitions approaches infinity. The expected value is also known as the expectation, mathematical expectation, average, mean value, mean.

More logically, the expected value of a discrete random variable is the probability-weighted average of all possible values. In other words, each possible value the random variable can assume is multiplied by its probability of occurrence, and the resultant are summed to produce the expected value. The same rule applies to a continuous random variable, except that an integral of the variable with respect to its probability density replaces the sum. The proper definition subsumes both of these and also works for distributions which are neither discrete nor continuous; the expected value of a random variable is the integral of the random variable with respect to its probability measure.

The expected value does not exist for random variables having some distributions with large "tails", such as the Cauchy distribution. For random variables such as these, the long-tails of the distribution put an end to the sum/integral from converging.

The expected value is a key attribute of how one characterizes a probability distribution; it is one type of place parameter. By contrast, the variance is a measure of dispersion of the possible values of the random variable around the expected value. The variance itself is defined in terms of two expectations: it is the expected value of the squared deviation of the variable's value from the variable's expected value.

The expected value plays significant roles in a diversity of contexts. In regression analysis, one needs a formula in terms of observed data that will give a "good" approximation of the parameter giving the effect of some clarifying variable upon a dependent variable. The formula will give different estimates using different samples of data, so the estimate it gives is itself a random variable. A procedure is normally considered good in this situation if it is an unbiased estimator that is, if the expected value of the estimate (the average value it would give over an arbitrarily large number of separate samples) can be shown to equal the true value of the preferred parameter.

In decision theory, and in particular in choice under ambiguity, an instrument is described as making the best choice in the context of deficient information. For risk neutral agents, the choice involves using the expected values of unsure quantities, while for risk average agents it involves maximizing the expected value of some objective function such as a von Neumann–Morgenstern utility function. One example of using expected value in reaching best possible decisions is the Gordon–Loeb model of information security investment. According to the model, one can wrap up that the amount a firm spends to protect information should generally be only a small fraction of the expected loss.

Key Terms and Formulas

Expectation

Let x be a simple random variable, that is, a discrete random variable taking on only finitely many possible values. Then expected value will be

$$E(x) = x_1 f(x_1) + x_2 f(x_2) + x_3 f(x_3) + \dots + x_n f(x_n) + \dots$$

$$E(x) = \sum_{i=1}^{\infty} x_i f(x_i),$$ provided it converges absolutely.

The sum converges absolutely if and only if $\sum |x| f(x)$ is finite

$E(x)$ is also called the mean of X and is usually denoted by the letter μ. It should be noted that $E(x)$ is the average value of the random variable X over a very large number of trials. So we can write it as

$E(x) = \sum x f(x)$ if X is discrete, and $\mu = E(x) = \int_{-\infty}^{\infty} x f(x) dx$ if x is continuous.

Provided the integral converges absolutely, i.e. $\int_{-\infty}^{\infty} |x| f(x) dx$ is finite.

Joint Probability Distribution

We shall now extend our concept of mathematical expectation to the case of two random variables X and Y with joint probability distribution $f(x, y)$.

Let X and Y be random variables with joint probability distribution $f(x, y)$. The mean, or expected value, of the random variable $g(X, Y)$ is

$\mu_g(X, Y) = E[g(X, Y)] = \sum_x \sum_y g(x, y) f(x, y)$ if X *and* Y are discrete,

and

$$\mu_g(X,Y)=E[g(X,Y)]=\int_{-\infty}^{\infty}\int_{-\infty}^{\infty}g(x,y)f(x,y)\,dx\,dy$$

if X and Y are continuous.

Variance of Random Variable

The mean, or expected value, of a random variable X is of special importance in statistics because it describes where the probability distribution is centered. By itself, however, the mean does not give an adequate description of the shape of the distribution. We also need to characterize the variability in the distribution.

Now, if $H(X)=(X-\mu)^2$, where μ is the population mean, then

$E(X-\mu)^2=\Sigma(x_i-\mu)^2 f(x)$

We call this expected value of the *variance* and denoted it by $Var(X)$ or σ^2.

That is, $\sigma^2=E(X-\mu)^2=E(X^2)-[E(X)]^2$

Properties of Variance

$Var(X)$ can not be negative.

$Var(a)=0$, where a is a constant.

$Var(aX)=a^2\,Var(X)$

$Var(aX+b)=a^2Var(X)$, where a and b are the constants.

Important Properties

$E(a)=a$

If X is a discrete r.v. and if a and b are constants, then $E(aX+b)=a\,E(X)+b.$

The expected value of the sum of any two random variables is equal to the sum of their expected values, i.e. $E(X+Y)=E(X)+E(Y)$

The expected value of the product of two independent r.v. is equal to the product of their expected values, i.e. $E(XY)=E(X)E(Y)$.

Covariance of Two random Variable

The covariance of two r.v. $X\ and\ Y$ is a numerical measure of the extent to which their values tend to increase or decrease together. It is denoted by σ_{xy} or $Cov(X,Y)$, and is defined as the expected value of the product $[X-E(X)][Y-E(Y)]$.

$$\boldsymbol{Cov(X,Y)=E(XY)-E(X)E(Y)}$$

if X and Y are independent

$$Var(X+Y)=Var(X)+Var(Y).$$

$$Var(X-Y)=Var(X)+Var(Y).$$

Correlation Coefficient of Random Variables

$$\tilde{n}_{XY}=\frac{\boldsymbol{Cov(X,Y)}}{\sqrt{\boldsymbol{Var(X)Var(Y)}}}$$

If $X\ and\ Y$ are independent r.v., then ρ_{XY} will be zero but zero correlation does not necessarily imply independence. The correlation coefficient has the following properties:

Correlation coefficient is unit less and symmetric in $X\ and\ Y, i.e., \rho_{XY}=\rho_{YX}$

Correlation coefficient remains unchanged if constants are added to the r.v. or if the r.v. are multiplied by constants having the same sign.

Correlation coefficient is a number between $-1\ and\ +1$ inclusive. Or we can write it

$$-1\le\rho\le1$$

Chebyshev's Inequality

If X is a r.v. having mean μ and variance $\sigma^2 > 0$, and k is any positive constant, then the probability that a value of X falls within k standard deviations of the mean is atleast $\left(1-\frac{1}{k^2}\right)$.

That is, $$P(\mu - k\sigma < X < \mu + k\sigma) \geq 1 - \frac{1}{k^2},$$

or equivalently, $$P\left[|x-\mu| \geq k\sigma\right] \leq \frac{1}{k^2}$$

Moment Generating Function

The moment generating function (m.g.f.) usually denoted by $M_0(t)$, of a random variable X about the origin if it exists, is defined as the expected value of the r.v. e^{iX}, where t is a real variable lying in a neighborhood of zero. That is

$$M_0(t) = E\left(e^{iX}\right) = \sum_{i=1}^{\infty} e^{tx_i} f(x_i), \quad if\ X\ is\ discrete,$$

$$= \int_{-\infty}^{\infty} e^{tx} f(x)\,dx, \quad if\ X\ is\ continuous.$$

CHAPTER 5

DISCRETE PROBABILITY DISTRIBUTION-I

INTRODUCTION

No matter whether a discrete probability distribution is represented graphically by a histogram, in tabular form, or by means of a formula, the behavior of a random variable is described. Often, the observations generated by different statistical experiments have the same general type of behavior. Consequently, discrete random variables associated with these experiments can be described by essentially the same probability distribution and therefore can be represented by a single formula. In fact, one needs only a handful of important probability distributions to describe many of the discrete random variables encountered in practice.

Such a handful of distributions describe several real-life random phenomena. For instance, in a study involving testing the effectiveness of a new drug, the number of cured patients among all the patients who use the drug approximately follows a binomial distribution. In an industrial example, when a sample of items selected from a batch of production is tested, the number of defective items in the sample usually can be modeled as a hypergeometric random variable. In a statistical quality control problem, the experimenter will signal a shift of the process mean when observational data exceed certain limits.

The number of samples required to produce a false alarm follows a geometric distribution which is a special case of the negative binomial distribution. On the other hand, the number of white cells from a fixed amount of an individual's blood sample is usually random and may be described by a Poisson distribution. In this chapter, we present these commonly used distributions with various examples.

Binomial Probability Distribution

Many experiments consist of repeated independent trials, each trial having only two possible complementary outcomes. For example, the two possible outcomes of a trial may be head and tail, success and failure, right and wrong, alive and dead, good and defective, infected and not infected and so forth. If the probability of each outcome remains the same throughout the trials, then such trials are called the Bernoulli trials and the experiment having n Bernoulli trials is called a binomial probability experiment if it possess the following four properties.

- The outcomes of each trial may be classified into one of two categories, conventionally called success (S) and failure (F). It is to be noted that the outcome of interest is called a success and the other, a failure.
- The probability of success, denoted by remains constant for all trials.
- The successive trials are all independent.
- The experiment is repeated a fixed number of times, say n.

When X denotes the number of successes in n trials of a binomial probability experiment, it is called a binomial random variable and its p.d. is called the *Binomial Probability Distribution*. The r.v. X can obviously take on anyone of the $(n+1)$ integer values $0,1,2,\ldots,n.$

When the binomial r.v. X assumes a value $x,$ the binomial p.d is given by

$$f(x)=P(X=x)=\binom{n}{x}p^x q^{n-x}, x=0,1,2,\ldots,n.$$

Where $q=1-p;$ the probability of failure on each trial. The binomial p.d. has two parameters n and p and is generally denoted by $b(x;n,p).$ The binomial probability distribution is appropriate when a random sample of size n is drawn with replacement from a finite population of size $N,$ or sampling is done from an infinite population.

The binomial p.d., which is the mostly widely used distribution in two-outcome situations, was discovered by the Swiss mathematician Jakob Bernoulli (1654-1704) whose main work on probability, the *Ars Conjectandi* (the art of conjecturing) was published posthumously in Basel in 1713.

Derivation of Binomial Probability Distribution

To derive a formula that gives the probability of successes in n trials for a binomial experiment, we proceed as follows:

The experiment has n trials, each of which may result in $S \, or \, F$. The sample space has 2^n possible sample points or outcomes, each outcome consisting of a sequence $\{a_1, a_2, \ldots, a_n\}$, where each a_i, is either $S \, or \, F$. We desire to find the probability of these outcomes according to the number of successes.

First, we consider the probability of zero success, i.e. $P(X=0)$. In case of zero success, every trial result in F and the event consists of a sequence of n F's, i.e. $\{F F \ldots F\}$. Because in each trial, $P(S)=p$ and $P(F)=q$ and trials are independent, so we apply the multiplicative law of probability for independent events and obtain

$$P(F F \ldots F) = P(F)P(F)\ldots P(F) \quad [n \, times]$$

$$= q^n$$

Since there is only one sequence of outcomes of n trials resulting in $F's$, therefore

$$P(X=0) = q^n$$

Next, we consider the probability of one success, i.e. $P(X=1)$. In this case, one trial results in S and the remaining $(n-1)$ trials result in $F's$.

The event consisting of one S and $(n-1)F's$ can occur in several different sequences. One such sequence is $\{S F F \ldots F\}$ and the probability for this sequence in pq^{n-1}. Another possible sequence is $\{FFSF\ldots F\}$ and the probability for this sequence is the same as that of the first sequence. In other words, the probability for any possible sequence consisting of one S and $(n-1)F's$ can occur, is $\binom{n}{1}$. Therefore the probability of exactly one success for all possible sequences combined, is $P(X=1)=\binom{n}{1}pq^{n-1}$. The above argument may be repeated for $X=2,3,4,etc.$

Finally, we consider the general case, i.e. when $X=x$. The probability of a sequence that has exactly x successes and $(n-x)$ failures is $p^x q^{n-x}$ and there are $\binom{n}{x}$ different sequences in which x successes and $(n-x)$ failures can occur. Therefore the probability of x successes in n trials is

$$P(X=x)=\binom{n}{x}p^x q^{n-x}, \qquad for\, x=0,1,2,\ldots,n.$$

Thus we have obtained the formula for the binomial probability distribution having n trials and the probability p for success. The binomial probability distribution derives its name from the fact that the probabilities $\binom{n}{x}p^x q^{n-x}$, for $x=0,1,2,\ldots,n$ are the successive terms of the binomial expension of $(p+q)^n$. That is,

$$(q+p)^n=\Sigma\binom{n}{x}p^x q^{n-x}, where\, x=0,1,2,\ldots,n$$

$$=\binom{n}{0}q^n+\binom{n}{1}q^{n-1}p+\binom{n}{2}q^{n-2}p^2+\ldots+\binom{n}{n}p^n$$

$$=b(0;n,p)+b(1;n,p)+b(2;n,p)+\ldots+b(n;n,p).$$

The sum of probabilities, i.e.

$$\sum_{x=0}^{n}\binom{n}{x}p^x q^{n-x} \; or \sum_{x=0}^{n} b\,(x\,;n,p)=1$$

Because $q+p=1.$ This is a necessary condition for any probability distribution. It is to be noted that the probability of r or fewer successes is given by

$$P(X\le r)=\sum_{x=0}^{r}\binom{n}{x}p^x q^{n-x}, where\, P(X\le r)$$

is the cumulative binomial distribution of $X.$ There are tables for the cumulative probabilities $P(X\le r)=\sum_{x=0}^{r} b(x;n,p)$ for some values of n, p and $r.$

Example 1: A fair coin is tossed 5 times. Find the probabilities of obtaining various numbers of heads.

***Solution*:** Let us regard that the tossing of a coin as an experiment. Then we observe that

- Each toss of a coin (i.e. each trial) has two possible outcomes; heads (Successes) and tails (failure);
- The probability of a head (success) is and remains the same for successive tosses;
- The successive tosses of the coin are independent; and
- The coin is tossed 5 times.

***Solution*:** Let us regard that the tossing of a coin as an experiment. Then we observe that

- Each toss of a coin (i.e. each trial) has two possible outcomes; heads (Successes) and tails (failure);
- The probability of a head (success) is $p=\frac{1}{2}$ and remains the same for successive tosses;
- The successive tosses of the coin are independent; and
- The coin is tossed 5 times.

Therefore the r.v. X which denotes the number of heads (successes) has a binomial probability distribution with $p=\frac{1}{2}$ and $n=5$. The possible values of X are $0,1,2,3,4\, and\, 5$. Hence

$$P(no\, head)=P(X=0)=\binom{5}{0}\left(\frac{1}{2}\right)^0\left(\frac{1}{2}\right)^5=\frac{1}{32}$$

$$P(1\, head)=P(X=1)=\binom{5}{1}\left(\frac{1}{2}\right)^1\left(\frac{1}{2}\right)^4=\frac{5}{32}$$

$$P(2\, head)=P(X=2)=\binom{5}{2}\left(\frac{1}{2}\right)^2\left(\frac{1}{2}\right)^3=\frac{10}{32}$$

$$P(3\, head)=P(X=3)=\binom{5}{3}\left(\frac{1}{2}\right)^3\left(\frac{1}{2}\right)^2=\frac{10}{32}$$

$$P(4\, head)=P(X=4)=\binom{5}{4}\left(\frac{1}{2}\right)^4\left(\frac{1}{2}\right)^1=\frac{5}{32}$$

$$P(5\, head)=P(X=5)=\binom{5}{5}\left(\frac{1}{2}\right)^5\left(\frac{1}{2}\right)^0=\frac{1}{32}$$

These probabilities can also be obtained by expanding the binomial $\left(\frac{1}{2}+\frac{1}{2}\right)^5$. The binomial probability distribution for the number of heads obtained in 5 tosses of a fair coin is

x	0	1	2	3	4	5
$f(x)$	$\frac{1}{32}$	$\frac{5}{32}$	$\frac{10}{32}$	$\frac{10}{32}$	$\frac{5}{32}$	$\frac{1}{32}$

Example 2: An event has the probability $p=\frac{3}{8}$. Find the complete binomial distribution for $n=5$ trials.

Solution: Here $p = \frac{3}{8}$ so that $q = 1 - p = \frac{5}{8}$; and $n = 5$.

Hence the desired probabilities are the successive terms in the binomial expansion of $\left(\frac{5}{8} + \frac{3}{8}\right)^5$

$$\left[\left(\frac{5}{8}\right)^5 + \binom{5}{1}\left(\frac{5}{8}\right)^4\left(\frac{3}{8}\right) + \binom{5}{2}\left(\frac{5}{8}\right)^3\left(\frac{3}{8}\right)^2 + \binom{5}{3}\left(\frac{5}{8}\right)^2\left(\frac{3}{8}\right)^3 + \binom{5}{4}\left(\frac{5}{8}\right)^1\left(\frac{3}{8}\right)^4 + \binom{5}{5}\left(\frac{5}{8}\right)^0\left(\frac{3}{8}\right)^5\right]$$

i.e. $\frac{1}{(8)^5}\left[(5)^5 + 5(5)^4(3) + 10.(5)^3(3)^2 + 10(5)^2(3)^3 + 5(5)(3)^4(3)^5\right]$

i.e. $\frac{1}{32768}\left[3125 + 9375 + 11250 + 6750 + 2025 + 243\right]$

i.e. [üüüüüü#üüüüüüü#üüüüü + + +]

We can now write these probabilities in the form of a probability table as bellow:

x	0	1	2	3	4	5
$f(x)$	0.0954	0.2861	0.3433	0.2060	0.0618	0.0074

Example: Let X have a binomial distribution with $n = 4$ and $p = \frac{1}{3}$

Find $P(X = 1), P\left(X = \frac{3}{2}\right)$

$P(X = 3), P(X = 6)\, and\, P(X \leq 2)$.

Solution: The binomial probability distribution for $n = 4\, and\, p = \frac{1}{3}, is$

$$f(x) = \binom{4}{x}\left(\frac{1}{3}\right)^x\left(\frac{2}{3}\right)^{4-x} \quad for\, x = 0,1,2,3,4.$$

Now $P(X=1)=\binom{4}{1}\left(\frac{1}{3}\right)^1\left(\frac{2}{3}\right)^3=\frac{32}{81};$

$P\left(X=\frac{3}{2}\right)=f\left(\frac{3}{2}\right)=0;$ because a r.v. X with a binomial distribution takes only one of the integer values $0,1,2,3,\ldots,n.$

$P(X=3)=f(3)=\binom{4}{3}\left(\frac{1}{3}\right)^3\left(\frac{2}{3}\right)^1=\frac{8}{81};$

$P(X=6)=f(6)=0, because$ X can take only values $0,1,2,3,4.$

And $P(X\le 2)=\sum_{x=0}^{2} f(x)=f(0)+f(1)+f(2)$

$$=\binom{4}{0}\left(\frac{1}{3}\right)^0\left(\frac{2}{3}\right)^4+\binom{4}{1}\left(\frac{1}{3}\right)^1\left(\frac{2}{3}\right)^3+\binom{4}{2}\left(\frac{1}{3}\right)^2\left(\frac{2}{3}\right)^2+\binom{4}{3}\left(\frac{1}{3}\right)^1\left(\frac{2}{3}\right)^3+\binom{4}{4}\left(\frac{1}{3}\right)^0\left(\frac{2}{3}\right)^4$$

$$=\frac{16}{81}+\frac{32}{81}+\frac{24}{81}=\frac{72}{81}=\frac{8}{9}$$

Example: A and B play a game in which A's probability of winning is $\frac{2}{3}$. In a series of 8 games, what is the probability that A will win (i) exactly 4 games; (ii) at least 4 games; (iii) 6 or more games; (iv) From 3 to 6 games?

Solution: We observe that

- There are two possible outcomes, i.e. A will win or will not win the game.
- The probability of A's winning in each game is $p=\frac{2}{3}$.
- The successive games are independently won or lost; and
- There are 8 games

Therefore, the binomial probability distribution with $n=8$ and $p=\frac{2}{3}$ is appropriate.

Let X denote the number of games won by A. Then

$$P(X=4)=\binom{8}{4}\left(\frac{2}{3}\right)^4\left(\frac{1}{3}\right)^4=\frac{1120}{6561}=0.1707$$

$$P(X\geq 4)=1-P(X<4);\ (atleast\,4\,means\,4\,or\,more)$$

$$=1-\sum_{x=0}^{3}\binom{8}{x}\left(\frac{2}{3}\right)^x\left(\frac{1}{3}\right)^{8-x}$$

$$=1-\left[\left(\frac{1}{3}\right)^8+8\left(\frac{2}{3}\right)\left(\frac{1}{3}\right)^7+28\left(\frac{2}{3}\right)^2\left(\frac{1}{3}\right)^6+56\left(\frac{2}{3}\right)^3\left(\frac{1}{3}\right)^5\right]$$

$$=1-\frac{1}{6561}[1+16+112+448]$$

$$=1-\frac{577}{6561}=\frac{5984}{6561}=0.9121$$

$$P(X\geq 6)=\sum_{x=6}^{8}\binom{8}{x}\left(\frac{2}{3}\right)^x\left(\frac{1}{3}\right)^{8-x}$$

$$=\binom{8}{6}\left(\frac{2}{3}\right)^6\left(\frac{1}{3}\right)^2+\binom{8}{7}\left(\frac{2}{3}\right)^7\left(\frac{1}{3}\right)+\binom{8}{8}\left(\frac{2}{3}\right)^8$$

$$=\frac{64}{6561}[28+16+4]=\frac{64\times 48}{6561}=\frac{1024}{2187}=0.4682$$

$$P(3\leq X\leq 6)=\sum_{x=3}^{6}\binom{8}{x}\left(\frac{2}{3}\right)^x\left(\frac{1}{3}\right)^{8-x}$$

$$=\binom{8}{3}\left(\frac{2}{3}\right)^3\left(\frac{1}{3}\right)^5+\binom{8}{4}\left(\frac{2}{3}\right)^4\left(\frac{1}{3}\right)^4+\binom{8}{5}\left(\frac{2}{3}\right)^3\left(\frac{1}{3}\right)^5+\binom{8}{6}\left(\frac{2}{3}\right)^2\left(\frac{1}{3}\right)^6$$

$$=\frac{2^3}{(3)^8}[56+140+224+224]$$

$$=\frac{8\times 644}{6561}=\frac{5152}{6561}=0.7852$$

Where Does the Name *Binomial* Come From?

The binomial distribution derives its name from the fact that the $n + 1$ terms in the binomial expansion of $(q+p)^n$ correspond to the various values of $b(x;n,p)$ for $x=0,1,2,\ldots,n.$ That is,

$$(q+p)^n=\binom{n}{0}q^n+\binom{n}{1}pq^{n-1}+\binom{n}{2}p^2q^{n-2}+\ldots\binom{n}{n}p^n$$

$$=b(0;n,p)+b(1;n,p)+b(2;n,p)+\ldots+b(n;n,p).$$

Since $p+q=1,$ we see that

$$\sum_{x=0}^{n}b(x;n,p)=1,$$

a form that must hold for any probability distribution.

Normally, we are interested in problems where it is required to find $P(X<r)$

or $P(a\leq X\leq b)$. Binomial sums

$$B(r;n,p)=\sum_{x=0}^{r}b(x;n,p)$$

Example: The probability that a patient recovers from a rare blood disease is 0.4. If 15 people are known to have contracted this disease, what is the probability that (a) at least 10 survive, (b) from 3 to 8 survive, and (c) exactly 5 survive?

Solution: Let X be the number of people who survive.

$$P(X\geq 10)=1-P(X<10)=1-\sum_{x=0}^{9}b(x\;;15,0.4)$$

$$=1-0.9662=0.0338$$

$$P(3 \le X \le 8) = \sum_{x=3}^{8} b(x\ ;15,0.4) = \sum_{x=0}^{8} b(x\ ;15,0.4) - \sum_{x=0}^{2} b(x\ ;15,0.4)$$

$$= 0.9050 - 0.0271 = 0.8779$$

$$P(X = 5) = b(15\ ;15,0.4) = \sum_{x=0}^{5} b(x\ ;15,0.4) - \sum_{x=0}^{4} b(x\ ;15,0.4)$$

$$= 0.4032 - 0.2173 = 0.1859$$

***Example*:** A large chain retailer purchases a certain kind of electronic device from a manufacturer. The manufacturer indicates that the defective rate of the device is 3%.

(a) The inspector randomly picks 20 items from a shipment. What is the probability that there will be at least one defective item among these 20?

(b) Suppose that the retailer receives 10 shipments in a month and the inspector randomly tests 20 devices per shipment. What is the probability that there will be exactly 3 shipments each containing at least one defective device among the 20 that are selected and tested from the shipment?

***Solution*:** (a) Let X *be* the number of defective devices among the 20. Then X follows a

$b(x;20,0.03)$ distribution. Hence,

$$P(X \ge 1) = 1 - P(X = 0) = 1 - b(0;20,0.03)$$

$$= 1 - (0.03)^0 (1 - 0.03)^{20-0} = 0.4562.$$

(b) In this case, each shipment can either contain at least one defective item or not. Hence, testing of each shipment can be viewed as a Bernoulli trial with $p = 0.4562$ from part (a). Assuming independence from shipment to shipment and denoting by Y the number of shipments containing at least one defective item, Y follows another binomial distribution $b(y;10,0.4562)$. Therefore,

$$P(Y = 3) = \binom{10}{3} (0.4562)^3 (1 - 0.4562)^7 = 0.1602$$

Areas of Application

From above Examples, it should be clear that the binomial distribution finds applications in many scientific fields. An industrial engineer is keenly interested in the "proportion defective" in an industrial process. Often, quality control measures and sampling schemes for processes are based on the binomial distribution. This distribution applies to any industrial situation where an outcome of a process is dichotomous and the results of the process are independent, with the probability of success being constant from trial to trial. The binomial distribution is also used extensively for medical and military applications.

In both fields a success or failure result is important. For example, "cure" or "no cure" is important in pharmaceutical work, and "hit" or "miss" is often the interpretation of the result of firing a guided missile. Since the probability distribution of any binomial random variable depends only on the values assumed by the parameters n, p, and q, it would seem reasonable to assume that the mean and variance of a binomial random variable also depend on the values assumed by these parameters.

Example: The experience of a house agent indicates that he can provide suitable accommodation for 75 per cent of the clients who come to him. If on, a particular occasion, 6 clients approach him independently, calculate the probability that

- Less than 4 clients.
- Exactly 4 clients
- At least 5 clients, will get satisfactory accommodation.

Solution: We observed that

- There are two possible outcomes, i.e. each client will get or will not get or will not get accommodation
- On each occasion, probability of getting accommodation is $p = \frac{3}{4}$
- Clients approach the house agent independently
- There are 6 clients

Therefore the binomial probability distribution with $n = 6$ and $p = \frac{3}{4}$ is appropriate to calculate the desired probabilities.

Let X denote the number of clients who get satisfactory accommodation. Then we need to calculate

$i. P(X < 4), ii) P(X = 4). iii) P(X \geq 5)$

i. $$P(X < 4) = \sum_{x=0}^{3} \binom{6}{x} \left(\frac{3}{4}\right)^x \left(\frac{1}{4}\right)^{6-x}$$

$$= \left(\frac{1}{4}\right)^6 + \binom{6}{1}\left(\frac{3}{4}\right)\left(\frac{1}{4}\right)^5 + \binom{6}{2}\left(\frac{3}{4}\right)^2\left(\frac{1}{4}\right)^4 + \binom{6}{3}\left(\frac{3}{4}\right)^3\left(\frac{1}{4}\right)^3$$

$$= \left(\frac{1}{4}\right)^6 \left[1 + (6)(3) + (15)(9) + (20)(27)\right]$$

$$= \frac{694}{4096} = 0.169$$

ii. $$P(X = 4) = \binom{6}{4}\left(\frac{3}{4}\right)^4\left(\frac{1}{4}\right)^2 = \frac{1215}{4096} = 0.297$$

iii. $$P(X \geq 5) = \sum_{x=5}^{6} \binom{6}{x} \left(\frac{3}{4}\right)^x \left(\frac{1}{4}\right)^{6-x}$$

$$= \binom{6}{5}\left(\frac{3}{4}\right)^5\left(\frac{1}{4}\right) + \binom{6}{6}\left(\frac{3}{4}\right)^6$$

$$= \left(\frac{1}{4}\right)^6 \left[(6)(3)^5 + (3)^6\right] = \frac{2187}{4096} = 0.534$$

Binomial Frequency Distribution

If the binomial probability distribution is multiplied by N, the number of experiments or sets, the resulting distribution is known as the binomial frequency distribution. Thus the expected frequency of x successes in N experiments is $N.\binom{n}{x}p^x q^{n-x}$. It should be noted that the n independent trials constitute one experiment or one set.

***Example*:** Six dice are thrown 729 times. How many times do you expect at least three dice to show a five or a six?

***Solution*:** The probability of getting a 5 or a 6 with one die is $p=\frac{2}{6}$. Since 6 dice are thrown and there are 729 sets, the binomial frequency distribution is given by

$$N(q+p)^n = 729\left(\frac{2}{3}+\frac{1}{3}\right)^6$$

Hence the expected number of times at least 3 dice showing 5 or 6

$$= 729\left[\sum_{x=3}^{6}\binom{6}{x}\left(\frac{1}{3}\right)^x\left(\frac{2}{3}\right)^{6-x}\right]$$

$$= 729\left[\binom{6}{3}\left(\frac{1}{3}\right)^3\left(\frac{2}{3}\right)^3+\binom{6}{4}\left(\frac{1}{3}\right)^4\left(\frac{2}{3}\right)^2+\binom{6}{5}\left(\frac{1}{3}\right)^5\left(\frac{2}{3}\right)+\left(\frac{1}{3}\right)^6\right]$$

$$= \frac{729}{(3)^6}\left[160+60+12+1\right] = 233$$

Example: A certain event is believed to follow the binomial distribution. In 1024 samples of 5, the result was observed once 405 times and twice 270 times. Find p and q.

Solution: The first three terms in the expansion of the binomial frequency distribution of $N(p+q)^n$ corresponding to $x=0,1 and\, 2$ are $Nq^n, N\binom{n}{1}q^{n-1}p\, and\, N\binom{N}{2}q^{n-2}p^2$.

We are given $N=1024, n-5$ and the following information:

$$1024\binom{5}{1}q^{5-1}p=405,$$

$$1025\binom{5}{2}q^{5-2}p^2=270,,$$

Dividing the second equation by the first, we get

$$\frac{10q^3p^2}{5q4p}=\frac{270}{405}$$

$$\frac{2p}{q}=\frac{2}{3}$$

$$3p=q\,or\,3p=1-p\,or\,4p=1$$

$$p=\frac{1}{4}\,and\,q=\frac{3}{4}$$

Properties of the Binomial Probability Distribution

The properties of the binomial distribution include the mean number of success, the variance of the number of successes, measure of skewness and kurtosis, etc. and the shape of the distribution. Some of the properties are described below:

- Let X be a random variable with the binomial distribution $b(x;n,p)$,then its mean and variance are given by $\mu=np$ and $\sigma^2=npq$, respectively.

Proof: Since we know that Mean, $E(x)$

$$=\sum_{x=0}^{n}x\binom{n}{x}p^xq^{n-x},\quad where\,x=0,1,2,\ldots,n.$$

$$=0.q^n+1.\binom{n}{1}q^{n-1}p+2\binom{n}{2}q^{n-2}p^2+\ldots+np^n$$

$$= np\left[q^{n-1} + \binom{n-1}{1} q^{n-2} p + \binom{n-1}{2} q^{n-3} p^2 + \ldots + p^{n-1}\right]$$

$$= np(q+p)^{n-1}$$

$$= np, because\, q + p = 1.$$

Now, for variance $\sigma^2 = E[X-\mu]^2 = E(X^2) - [E(X)]^2$

But $E(X^2) = E[X(X-1)+X] = E[X(X-1)] + E(X)$

$$= E[X(X-1)] + np$$

Now

$$E[X(X-1)] = \sum_{x=0}^{n} x(x-1)\binom{n}{x} p^x q^{n-x}$$

$$= \sum_{x=0}^{n} x(x-1)\frac{n(n-1)(n-2)!}{x(x-1)(x-2)!(n-x)!} p^2 . p^{x-2} q^{n-x}$$

$$= n(n-1)p^2 \sum_{x=2}^{n} \frac{(n-2)!}{(x-2)!(n-x)!} p^{x-2} q^{n-x}$$

x starts at 2 since x = 0, 1 add nothing to the sum

The term $(n-x)$ may be written as $[(n-2)-(x-2)]$

Substituting $y = x-2$ and $m = n-2$ in the summation, we obtain

$$E[X(X-1)] = n(n-1)p^2 \sum_{y=0}^{m} \frac{m!}{y!(m-y)!} p^y q^{m-y}$$

$$= n(n-1)p^2 \sum_{y=0}^{m} \binom{m}{y} p^y q^{m-y}$$

$$= n(n-1)p^2$$

Thus $\sigma^2 = E(X^2) - [E(x)]^2$

$$= E\left[X(X-1)\right] + E(X) - \left[E(X)\right]^2$$

$$= n(n-1)p^2 + np - (np)^2$$

$$= n^2p^2 - np^2 + np - n^2p^2$$

$$= np - np^2$$

$$= np(1-p)$$

$$= npq$$

$$\sigma = \sqrt{npq}$$

Hence the variance of the number of successes is npq, and the standard deviation is $\sqrt{npq}$.

Moments

Higher moments of the distribution are found as below:

By definition, the moments about the origin are given by the relation

$$\mu_r' = E\left(X^r\right)$$

Now $\mu_1' = E(X) = np$

$$\mu_2' = E\left(X^2\right) = n(n-1)p^2 + np$$

$$\mu_3' = E\left(X^3\right)$$

But $X^3 = X(x-1)(X-2) + 3X(X-1) + X$

$$E\left(X^3\right) = E\left[X(X-1)(X-2)\right] + 3E\left[X(X-1)\right] + E(X)$$

Now

$$E(X) = \sum_{x=0}^{n} x \binom{n}{x} p^x q^{n-x} = np$$

$$3E\left[X(X-1)\right]=3\sum_{x=0}^{n}x(x-1)\binom{n}{x}p^{x}q^{n-x}$$

$$=3n(n-1)p^{2}$$

$$E\left[X(X-1)(X-2)\right]=\sum_{x=0}^{n}x(x-1)\binom{n}{x}p^{x}q^{n-x}$$

But $x(x-1)(x-2)\binom{n}{x}=x(x-1)(x-2)\dfrac{n(n-1)(n-2)(n-3)!}{x(x-1)(x-2)(x-3)!(n-x)!}$

$$=n(n-1)(n-2)\binom{n-3}{x-3}$$

$$E\left[X(X-1)(X-2)\right]=n(n-1)(n-2)\sum_{x=3}^{n}\binom{n-3}{x-3}p^{3}q^{n-x}$$

(X starts at 3 since x = 0, 1, 2 add nothing to the summation)

Substituting $y=x-3$ and $m=n-3$ in the summation, we have

$$e\,(X(X-1))(X-2)=n(n-1)(n-2)\,p^{3}\sum_{y=0}^{m}\binom{m}{y}p^{y}\,q^{m-y}$$

$$=n(n-1)(n-2)\,p^{3}$$

Thus $\mu_3' = n(n-1)(n-2)\,p^{3} + 3n(n-1)\,p^{2} + np$

Again $\mu_4' = E\left(X^{4}\right)$

$$=\sum_{x=0}^{n}x^{4}\binom{n}{x}p^{x}q^{n-x}$$

Writing x^{4} as $x(x-1)(x-2)(x-3)+6x(x-1)(x-2)+7x(x-1)+x$
and proceeding above, we get

$$\mu_4' = n(n-1)(n-2)(n-3)\,p^{4} + 6n(n-1)(n-2)\,p^{3} + 7n(n-1)\,p^{2} + np$$

Therefore $\mu_3 = \mu_3' - 3\mu_1'\mu_2' + 2\left(\mu_1'\right)^{3}$

$$=\left[n(n-1)(n-2)p^3+3n(n-1)p^2+np\right]-3np\left[n^2p^2-np^2+np\right]+2n^3p^3$$

$$=np\left[1-3p+2p^2\right]=np(1-p)(1-2p)$$

$$=npq(q-p);$$

And

$$\mu_4=\mu_4'-4\mu_1'\mu_3'+6\left(\mu_1'\right)^2\left(\mu_2'\right)-3\left(\mu_1'\right)^4.$$

Substitution and simplification give

$$\mu_4=n^4\left[p^4-q^4\right]+n^3\left(-6p^4+6p^3+6p^3-6p^4\right)$$

$$+n^2\left(11p^4-18p^3+7p^2-4p^2+12p^3-8p^4\right)+n\left(-6p^4+12p^3-7p^2+p\right)$$

$$=3n^2p^2(1-p)^2+np(1-p)\left(1-6p+6p^2\right)$$

$$=3n^2p^2q^2+npq(1-6pq)$$

$$=npq\left(1+3(n-2)pq\right)$$

$$\beta_1=\frac{\mu_3^2}{\mu_2^3}=\frac{\left[npq(p-q)\right]^2}{(npq)^3}=\frac{(q-p)^2}{npq}=\frac{(1-2p)^2}{npq}$$

$$\beta_2=\frac{\mu_1}{\mu_2^2}=\frac{3n^2p^2q^2+npq(1-6pq)}{(npq)^2}=3+\frac{1-6pq}{npq}$$

Hence moment coefficient of skewness and kurtosis are respectively

$$\frac{1-2p}{\sqrt{npq}}\ and\ 3+\frac{1-6pq}{npq}$$

Example: In the B.A. examination, 24 candidates offered statistics. If the probability of passing the subject be $\frac{1}{3}$, find the mean and dispersion of the distribution.

Solution: Here $n=24\ and\ p=\frac{1}{3}$ so that $q=1-p=\frac{2}{3}$

Hence $mean = np = 24 \times \frac{1}{3} = 8;$ and

$$\sigma^2 = npq = 24 \times \frac{1}{3} \times \frac{2}{3} = \frac{16}{3} = 5.33$$

Fitting a Binomial Distribution to Observed Data

Fitting a binomial distribution to a given frequency distribution consists in i) estimating the values of the two parameters that completely determine a binomial distribution, and ii) calculating the probabilities as well as the expected frequencies for

Assuming that the given frequency distribution has the characteristics of the fitted theoretical binomial distribution, we calculate the mean of the observed frequency distribution, and taking it as the estimate of we equate it to its theoretical value, i.e. where is usually taken as the largest x value given. Having found the value of we compute the expected frequencies. The procedure is illustrated by the following example.

***Example*:** Fit a binomial distribution to the following data, obtained by tossing a biased coin 5 times:

No.of Heads	0	1	2	3	4	5	Total
Frequency	12	56	74	39	18	1	200

To fit a binomial distribution, we need to find n and p. Here, $n = 5$ the largest x-value. To find p, we use the relationship $\overline{x} = np$.

Now, $\overline{x} = \frac{\Sigma f_i x_i}{\Sigma f_i} = \frac{(0\times12)+(1\times56)+(2\times74)+(3\times39)+(4\times18)+(5\times1)}{200} = \frac{398}{200} = 1.99$

Using the relation $\overline{x} = np$, we get $5p = 1.99 \, or \, p = 0.398$

Letting r.v. X represent the number of heads, gives the fitted binomial distribution as

$$b(x;5,0.398) = \binom{5}{x}(0.398)^x(0.602)^{5-x}$$

Now the probabilities and frequencies are calculated as below

No. of Heads	Probability $f(x)$	Expected Frequency
0	$\binom{5}{0} q^5 = 0.07907$	15.8
1	$\binom{5}{1} q^4 p = 0.26136$	52.5
2	$\binom{5}{1} q^3 p^2 = 0.34559$	69.1
3	$\binom{5}{1} q^2 p^3 = 0.22847$	45.7
4	$\binom{5}{1} q p^4 = 0.07553$	15.1
5	$\binom{5}{1} p^5 = 0.00998$	2.0
Total	1.0000	200

The expected frequencies are obtained by multiplying each of the probabilities by 200. The probabilities can also be calculated, using the binomial recurrence formula:

$$P(X = x) = \frac{n - x + 1}{x} . \frac{p}{q} P(X = x - 1)$$

Moment Generating and Cumulant Generating Functions of the Binomial Distribution

The m.g.f. of the binomial probability distribution $b(x; n, p)$ is derived as below

$$M_0(t) = E(e^{tX})$$

$$= \sum_{x=0}^{n} e^{tX} \binom{n}{x} p^x q^{n-x}$$

$$= \sum_{x=0}^{n} \binom{n}{x} \left(pe^t\right) q^{n-x} = \left(q + pe^t\right)^n$$

The expansion of this binomial is purely algebraic and need not be interpreted in terms of probabilities. We get the moments by differentiating $M_0(t)$ once, twice, etc. with respect to t and putting $t = 0$. Thus,

$$\mu_1' = E(X) = \left[\frac{d}{dt}\left(q + pe^t\right)^n\right]_{t=0}$$

$$= \left[npe^t\left(q + pe^t\right)^{n-1}\right]_{t=0}$$

$$\mu_2' = E\left(X^2\right) = \left[\frac{d^2}{dt^2}\left(q + pe^t\right)^n\right]_{t=0}$$

$$= \left[npe^t\left(q + pe^t\right)^{n-1}\right]_{t=0} + \left[n(n-1)p^2e^{2t}\left(q + pe^t\right)^{n-2}\right]_{t=0}$$

$$= np + n(n-1)p^2$$

$$\mu_2 = \mu_2' - \left(\mu_1'\right)^2 = npq$$

In a similar way, the higher moments are obtained.

The cumulant generating function (c.f.) is given by

$$k(t) = \log_e M_0(t)$$

$$= \log_e\left[q + pe^t\right]^n = n\log_e\left[q + pe^t\right]$$

$$= n\,log_e\left[q + p\left(1 + t + \frac{t^2}{2!} + \frac{t^3}{3!} + \ldots\right)\right]$$

$$= n\log_e\left[1 + \left(pt + \frac{pt^2}{2!} + \frac{pt^3}{3!} + \ldots\right)\right] \quad (\therefore q = 1 - p)$$

Expanding the log and comparing powers in t, we find the first four cumulants to be

$$k_1 = \mu_1' = np;$$

$$k_2 = \mu_2 = np(1-p) = npq;$$

$$k_3 = \mu_3 = np(1-p)(1-2p) = npq(q-p);$$

$$k_4 = np(1-p)(1-6p+6p^2) = npq(1-6pq).$$

Hypergeometric Probability Distribution

There are many experiments in which the condition of independence is violated and the probability of success does not remain constant for all trials. Such experiments are called hypergeometric experiments. In other words, a hypergeometric experiments has the following properties.

- The outcomes of each trial may be classified into one of two categories, success and failure.
- The probability of success changes on each trial.
- The successive trials are dependent.
- The experiment is repeated a fixed number of times.

The number of successes, X in a hypergeometric experiment is called a hypergeometric r.v. and its probability distribution is called the hypergeometric distribution. When the hypergeometric r.v. X assumes a value x, the hypergeometric p. d is given by the formula

$$P(X=x) = \frac{\binom{k}{x}\binom{N-k}{n-x}}{\binom{N}{n}},$$

$$\textit{for } x \textit{ such that } x = 0,1,2,\ldots,n \textit{ and } x = 0,1,2,\ldots,k$$

where $N =$ number of units in the set or population, (a positive integer)
$n =$ number of units in the subset or sample (a positive integer) and
$k =$ number of successes in the set or population (a non-negative integer less than or equal to N)

The hypergeometric p.d. has three parameters $(N, n\, and\, k,$ or $N, n\, and\, p = \frac{k}{N})$, and is generally denoted by $h(x; N, n, k)$. The hypergeometric p.d. is appropriate when

- A random sample of size n is drawn without replacement from a finite population of N units;
- k of the units are of one kind (classified as success) and the remaining $N-k$ of another kind (classifies as failure).

The simplest way to view the difference between the binomial distribution and the hypergeometric distribution is to the way the sampling is done. The types of applications for the hypergeometric are same to those for the binomial distribution. We are concerned in computing probabilities for the number of observations that fall into a particular type. But in the case of the binomial distribution, independence among trials is required. As a result, if that distribution is applied to say sampling from a lot of items (deck of cards, batch of production items), the sampling must be done with replacement of each item after it is observed. On the other hand, the hypergeometric distribution does not require independence and is based on sampling done without replacement.

If we wish to find the probability of observing 3 red cards in 5 draws from an ordinary deck of 52 playing cards, the binomial distribution does not apply unless each card is replaced and the deck reshuffled before the next draw is made. To solve the problem of sampling without replacement, let us restate the problem. If 5 cards are drawn at random, we are interested in the probability of selecting 3 red cards from the 26 available in the deck and 2 black cards from the 26 available in the deck. There are $\binom{26}{3}$ ways of selecting 3 red cards, and for each of these ways we can choose 2 black cards in $\binom{26}{2}$ ways. Therefore, the total number of ways to select 3 red and 2 black cards in 5 draws is the product $\binom{26}{3}\binom{26}{2}$. The total number of ways to select any 5 cards from the 52 that are available is

$\binom{52}{5}$. Hence, the probability of selecting 5 cards without replacement of which 3 are red and 2 are black is given by $\dfrac{\binom{26}{3}\binom{26}{2}}{\binom{52}{5}} = 0.3251$.

In general, we are interested in the probability of selecting x successes from the k items labeled successes and $n-x$ failures from the $N-k$ items labeled failures when a random sample of size n is selected from N items. This is known as a hypergeometric experiment. The number X of successes of a hypergeometric experiment is called a hypergeometric random variable. Accordingly, the probability distribution of the hypergeometric variable is called the hypergeometric distribution, and its values are denoted by $h(x;N,n,k)$, since they depend on the number of successes k in the set N from which we select n items.

Derivation of Hypergeometric Distribution

Suppose a set contains N elements, out of which k are classified as success and $N-k$ are classified as failure; the two classes being mutually exclusive and exhaustive, and we select a subset of n elements $(n \le N)$ from the set without replacement. Then the total number of ways in which a subset of n elements can be chosen from N is $\binom{N}{n}$.

Let X denote the number of successes and let $X = x$ if and only if we choose x successes from k successes and $n-x$ failures from $N-k$ failures in the set. Then the number of ways in which x successes and $n-x$ failures can be chosen, is $\binom{k}{x}\binom{N-k}{n-x}$.

Hence the probability that x of the n elements are successes, is

$$P(X=x)=\frac{\binom{k}{x}\binom{N-k}{n-x}}{\binom{N}{n}},$$

For x such that $0 \le x \le n \, and \, 0 \le x \le k.$

Which is the required formula for the hypergeometric probabilities.

Now to prove that the sum of the hypergeometric probabilities is one, we use the following useful mathematical result or formula.

$$\sum_{j=0}^{m}\binom{a}{j}\binom{b}{m-j}=\binom{a+b}{m}$$

Derived by expanding both sides of

$$(1+x)^a(1+x)^b=(1+x)^{a+b}$$

and then equating the coefficients of x^n.

Thus,

$$\sum_{x=0}^{n}P(X=x)=\frac{1}{\binom{N}{n}}\sum_{x=0}^{n}\binom{k}{x}\binom{N-k}{n-x}$$

$$=\frac{1}{\binom{N}{n}}\binom{N-k+k}{n-x+x}=\frac{1}{\binom{N}{n}}\binom{N}{n}=1$$

The hypergeometric distribution derives its name from the fact that probability generating function may be put in the form of a hypergeometric series.

***Example*:** An urn contains 4 red balls and 6 black balls. A sample of 4 balls is selected from the urn without replacement. Let X be the number of red balls contained in the sample, then find the probability distribution for X.

Solution: Here X is a hypergeometric r.v. because

- The results of each draw may be classified as either red (success) or black (failure),
- The probability of success changes on each draw,
- The successive draws are dependent as the selection is made without replacement,
- The drawing is repeated a fixed number of times $(n=4)$.

Now, $N=4+6=10,\ k=4, n=4$ and the possible values of X are $0,1,2,3\,and\,4.$ Therefore, the probabilities of theses possible outcomes are

$$P(X=0)=h(0;10,4,4)=\frac{\binom{4}{0}\binom{6}{4}}{\binom{10}{4}}=\frac{15}{210}$$

$$P(X=1)=h(1;10,4,4)=\frac{\binom{4}{1}\binom{6}{3}}{\binom{10}{4}}=\frac{80}{210}$$

$$P(X=2)=h(2;10,4,4)=\frac{\binom{4}{2}\binom{6}{2}}{\binom{10}{4}}=\frac{90}{210}$$

$$P(X=3)=h(3;10,4,4)=\frac{\binom{4}{3}\binom{6}{1}}{\binom{10}{4}}=\frac{24}{210}$$

$$P(X=4)=h(4;10,4,4)=\frac{\binom{4}{4}\binom{6}{0}}{\binom{10}{4}}=\frac{1}{210}$$

Hence the hypergeometric p.d. of X is as follows:

x	0	1	2	3	4
$h(x;10,4,4)$	$\frac{15}{210}$	$\frac{80}{210}$	$\frac{90}{210}$	$\frac{24}{210}$	$\frac{1}{210}$

***Example*:** The names of 5 men and 5 women are written on slips of paper and placed in a hat. Four names are drawn. What is the probability that 2 are men and 2 are women?

***Solution*:** Let X denote the number of men. Then

$N=5+5=10$ names to be drawn from:

$n=4,$ and (here possible values of $x\,are\,0,1,2,3,4\,i.e.n)$

$k=5$

Hence the hypergeometric distribution is

$$h(x;10,4,5)=\frac{\binom{5}{x}\binom{5}{4-x}}{\binom{10}{4}}=\frac{1}{210}$$

and the required probability i.e. $P(X=2)$ is

$$h(2;10,4,5)=\frac{\binom{5}{2}\binom{5}{2}}{\binom{10}{4}}=\frac{10}{210}$$

***Example*:** What is the probability that a poker hand will contain exactly 2 aces?

Solution: Let us regard the 4 aces as success and the 48 non aces to be failure. Then we have

$N = 52, n = 5,$ (number of cards in a poker hand)

$k = 4\, and\, x = 2.($ here, possible values of x are $0,1,2,3\, and\, 4\, i.e.k)$

Hence the probability that a poker hand will contain exactly 2 aces is

$$P(X=2) = h(2;52,5,4) = \frac{\binom{4}{2}\binom{48}{3}}{\binom{52}{5}} = \frac{2162}{54145} = 0.0399.$$

Example: A particular part that is used as an injection device is sold in lots of 10. The producer deems a lot acceptable if no more than one defective is in the lot. A sampling plan involves random sampling and testing 3 of the parts out of 10. If none of the 3 is defective, the lot is accepted. Comment on the utility of this plan.

Solution: Let us assume that the lot is truly unacceptable (i.e., that 2 out of 10 parts are defective). The probability that the sampling plan finds the lot acceptable is

$$P(X=0) = \frac{\binom{2}{0}\binom{8}{3}}{\binom{10}{3}} = 0.467$$

Thus, if the lot is truly unacceptable, with 2 defective parts, this sampling plan will allow acceptance roughly 47 % of the time. As a result, this plan should be considered fairly.

Example: Lots of 40 components each are deemed unacceptable if they contain 3 or more defective. The procedure for sampling a lot is to select 5 components at random and to reject the lot if a defective is found. What is the probability that exactly 1 defective is found in the sample if there are 3 defective in the entire lot?

Solution: Using the hypergeometric distribution with $n = 5, N = 40, k = 3\, and\, x = 1.$ we find the probability of obtaining 1 defective to be

$$h(1;40,5,3)=\frac{\binom{3}{1}\binom{37}{4}}{\binom{40}{5}}=0.3011$$

Once again, this plan is not desirable since it detects a bad lot (3 defective) only about 30% of the time.

Properties of Hypergeometric Distribution.

The important properties of the hypergeometric probability distribution are given here.

- The mean and variance of the Hypergeometric probability distribution are $\mu = np$ and $\sigma^2 = npq\frac{N-n}{N-1}$, $where\ p=\frac{k}{N} and\ q=\frac{N-k}{N}$.

Proof: Let X have the Hypergeometric probability distribution given by

$$h(x;N,n,k)=\frac{\binom{k}{x}\binom{N-k}{n-x}}{\binom{N}{n}}, for\ x\ such\ that\ 0\le x\le n\ and\ 0\le x\le k.$$

Then the mean μ is given by

$$\mu = E(X)$$

$$=\sum_{x=0}^{n}\frac{x.\binom{k}{x}\binom{N-k}{n-x}}{\binom{N}{n}}$$

$$=\sum_{x=0}^{n}\frac{x.k(k-1)!\binom{N-k}{n-x}}{x.(x-1)!(k-x)!\binom{N}{n}}=\sum_{x=0}^{n}\frac{k.\binom{k-1}{x-1}\binom{N-k}{n-x}}{\binom{N}{n}}$$

Let $y = x - 1$,

Then,

$$\mu = \frac{k}{\binom{N}{n}} \sum_{y=0}^{n-1} \binom{k-1}{y} \binom{N-k}{n-1-y} [when\, x = 1, y = 0\, and\; when\, x = n, y = n-1]$$

$$= \frac{k}{\binom{N}{n}} \binom{N-1}{n-1} \left[\sum_{j=0}^{m} \binom{a}{j} \binom{b}{m-j} = \binom{a+b}{m} \right]$$

$$= \frac{kn(n-1)!(N-n)!}{N(N-1)!} \cdot \frac{(N-1)!}{(n-1)!(N-n)!}$$

$$= \frac{nk}{N} = np, where\; p = \frac{k}{N}.$$

Thus the mean of the hypergeometric distribution is the same as that of the binomial distribution.

By definition, the variance σ^2 is given by

$$\sigma^2 = E[X-\mu]^2 = E(X^2) - \mu^2$$

Now $E(X^2) = E[X(X-1) + X] = E(X) + E[X(X-1)]$

$$= \sum_{x=0}^{n} x.h(x; N, n, k) + \sum_{x=0}^{n} x(x-1) h(x; N, n, k)$$

$$= \frac{nk}{N} + \frac{\sum_{x=0}^{n} x(x-1) \binom{k}{x} \binom{N-k}{n-x}}{\binom{N}{n}}$$

$$= \frac{nk}{N} + \frac{k(k-1)\sum_{x=2}^{n} \binom{k-2}{x-2}\binom{N-k}{n-x}}{\binom{N}{n}}$$

Let $y = x - 2, then$

$$E(X^2) = \frac{nk}{N} + \frac{k(k-1)\sum_{y=0}^{n-2} \binom{k-2}{y}\binom{N-k}{n-2-y}}{\binom{N}{n}} \quad [when\, x = 2, y = 0,$$

$$and\ when\, x = n, y = n-2\ ]$$

$$= \frac{nk}{N} + \frac{k(k-1)\binom{N-2}{n-2}}{\binom{N}{n}} = \frac{nk}{N} + \frac{k(k-1).n(n-1)}{N(N-1)}$$

$$\sigma^2 = E(X^2) - \mu^2$$

$$= \frac{nk}{N} + \frac{k(k-1)n(n-1)}{N(N-1)} - \left(\frac{nk}{N}\right)^2$$

$$= \frac{nk(N-k)}{N^2} \cdot \frac{N-n}{N-1}$$

$$= npq \cdot \frac{N-n}{N-1}, where\ p = \frac{k}{N} and\ q = \frac{N-k}{N}.$$

Proceeding along the same steps as above, we may obtain higher moments, but they are rather complicated.

- If N becomes indefinitely large, the hypergeometric probability distribution tends to the binomial probability distribution.

Proof: Let $p = \frac{k}{N}$. Then $k = Np\ and\ N - k = N(1-p) = Nq.$

Substituting these values in the hypergeometric formula, we get

$$h(x;N,n,k)=\frac{\binom{Np}{x}\binom{Nq}{n-x}}{\binom{N}{n}}$$

$$=\frac{n!(Np)!(Nq)!(N-n)!}{x!(Np-x)!(n-x)!(Nq-n+x)!N!}$$

$$=\binom{n}{x}\frac{(Np)!(Nq)!(N-n)!}{(Np-x)!(n-x)!(Nq-n+x)!N!}$$

Applying Stirling's approximation $\left(n!\cong e^{-n}.n^{n}\sqrt{2\pi n}\right)$ to all factorial terms and simplifying, we get

$$h(x;N,n,k)\cong\binom{n}{x}\frac{p^{Np+\frac{1}{2}}.q^{Nq+\frac{1}{2}}\left(1-\frac{n}{N}\right)^{n-n+\frac{1}{2}}}{\left(p-\frac{x}{N}\right)^{Np-x+\frac{1}{2}}\left(q-\frac{n-x}{N}\right)^{Nq-n+x+\frac{1}{2}}}$$

Now, if N is allowed to become indefinitely large, then $\frac{x}{N},\frac{n-x}{N}$ and $\frac{n}{N}$ each approaches zero. Therefore

$$h(x;N,n,k)\cong\binom{n}{x}p^{x}q^{n-x}=b(x;n,p).$$

Example: Lots of 40 components each are deemed unacceptable if they contain 3 or more defective. The procedure for sampling a lot is to select 5 components at random and to reject the lot if a defective is found. What is the probability that exactly 1 defective is found in the sample if there are 3 defective in the entire lot. Find mean and variance of the random variable and then use Chebyshev's theorem to interpret the interval $\mu\pm 2\sigma$.

Solution: Sine we already saw that it was a hypergeometric experiment with $N = 40, n = 5, and\ k = 3$ so by mean and variance formulas, we have

$$\mu = \frac{(5)(3)}{40} = \frac{3}{8} = 0.375,$$

And

$$\sigma^2 = \left(\frac{40-5}{39}\right)(5)\left(\frac{3}{40}\right)\left(1-\frac{3}{40}\right) = 0.3113$$

Taking square root of $0.3113,$ we find that $\sigma = 0.558.$ Hence the required interval is $0.375 \pm (2)(0.558)$ or from $-0.741 t0 1.491.$ Chebyshev's theorem states that the number of defectives obtained when 5 components are selected at random from a lot of 40 components of which 3 are defective has a probability of at least $\frac{3}{4}$ of falling between $-0.741 and 1.491.$ That is, at least three fourths of the time, the 5 components include fewer than 2 defectives.

Relationship to the Binomial Distribution

In this chapter, we discussed several important discrete distributions that have wide applicability. Many of these distributions relate nicely to each other. The beginning student should gain a clear understanding of these relationships. There is an interesting relationship between the hypergeometric and the binomial distribution. As one might expect, if *n* is small compared to *N*, the nature of the *N* items changes very little in each draw. So a binomial distribution can be used to approximate the hypergeometriciribution when *n* is small

Areas of Application

It should be clear that the binomial distribution finds applications in many scientific fields. An industrial engineer is keenly interested in the "proportion defective" in an industrial process. Often, quality control measures and sampling schemes for processes are based on the binomial distribution. This distribution applies to any industrial situation where an outcome of a process is dichotomous and the results of the process are independent, with the probability of success being constant from trial to trial. The binomial distribution is also used extensively for medical and military applications.

In both fields, a success of failure result is important. For example, "cure" or "no cure" is important in pharmaceutical work, and "hit" or "miss" is often the interpretation of the result of firing a guided missile. Since the probability distribution of any binomial random variable depends only on the values assumed by the parameters *n*, *p*, and *q*, it would seem reasonable to assume that the mean and variance of a binomial random variable also depend on the values assumed by these parameters.

Binomial Frequency Distribution

If the binomial probability distribution is multiplied by N, the number of experiments or sets, the resulting distribution is known as the binomial frequency distribution. Thus the expected frequency of x successes in N experiments is $N.\binom{n}{x}p^x q^{n-x}.$ It should be noted that the n independent trials constitute one experiment or one set.

Properties of the Binomial Probability Distribution:

The properties of the binomial distribution include the mean number of success, the variance of the number of successes, measure of skewness and kurtosis, etc. and the shape of the distribution. Some of the properties are described below:

1. *Mean and Variance*

Let X be a random variable with the binomial distribution $b(x;n,p)$. Then, its mean and variance are given by $\mu = np$ and $\sigma^2 = npq$ respectively.

2. *Moments*

$\mu_r^{'} = E(X^r)$

Now $\mu_1' = E(X) = np$

$\mu_2' = E(X^2) = n(n-1)p^2 + np$

$$\mu_3' = E(X^3) = n(n-1)(n-2)p^3 + 3n(n-1)p^2 + np$$

Again $\mu_4' = E(X^4)$

$$\mu_4 = \mu_4' - 4\mu_1'\mu_3' + 6(\mu_1')^2(\mu_2') - 3(\mu_1')^4.$$

Substitution and simplification give

$$\mu_4 = npq(1 + 3(n-2)pq)$$

$$\beta_1 = \frac{\mu_3^2}{\mu_2^3} = \frac{[npq(p-q)]^2}{(npq)^3} = \frac{(q-p)^2}{npq} = \frac{(1-2p)^2}{npq}$$

$$\beta_2 = \frac{\mu_1}{\mu_2^2} = \frac{3n^2p^2q^2 + npq(1-6pq)}{(npq)^2} = 3 + \frac{1-6pq}{npq}$$

Hence moment coefficient of skewness and kurtosis are respectively

$\dfrac{1-2p}{\sqrt{npq}}$ and $3 + \dfrac{1-6pq}{npq}$

Moment Generating and Cumulant Generating Functions of the Binomial Distribution.

The m.g.f. of the binomial probability distribution $b(x;n,p)$ is derived as below

$$M_0(t) = E(e^{tX})$$

$$= \sum_{x=0}^{n} e^{tX} \binom{n}{x} p^x q^{n-x}$$

$$= \sum_{x=0}^{n} \binom{n}{x} (pe^t) q^{n-x} = (q + pe^t)^n.$$

The expansion of this binomial is purely algebraic and need not be interpreted in terms of probabilities. We get the moments by differentiating $M_0(t)$ once, twice, etc. with respect to t and putting $t = 0$. Thus

$$\mu_1' = E(X) = \left[\frac{d}{dt}\left(q + pe^t\right)^n\right]_{t=0}$$

$$= \left[npe^t\left(q + pe^t\right)^{n-1}\right]_{t=0}$$

$$\mu_2' = E\left(X^2\right) = \left[\frac{d^2}{dt^2}\left(q + pe^t\right)^n\right]_{t=0}$$

$$= \left[npe^t\left(q + pe^t\right)^{n-1}\right]_{t=0} + \left[n(n-1)p^2e^{2t}\left(q + pe^t\right)^{n-2}\right]_{t=0}$$

$$= np + n(n-1)p^2$$

$$\mu_2 = \mu_2' - \left(\mu_1'\right)^2 = npq$$

In a similar way, the higher moments are obtained.

The cumulant generating function (c.f.) is given by

$$k(t) = \log_e M_0(t)$$

$$= \log_e\left[q + pe^t\right]^n = n\log_e\left[q + pe^t\right]$$

$$= n\,log_e\left[q + p\left(1 + t + \frac{t^2}{2\,!} + \frac{t^3}{3\,!} + \ldots\right)\right]$$

$$= n\log_e\left[1 + \left(pt + \frac{pt^2}{2\,!} + \frac{pt^3}{3\,!} + \ldots\right)\right] (\because q = 1 - p)$$

Expanding the log and comparing powers in t, we find the first four cumulants to be

$$k_1 = \mu_1' = np;$$

$$k_2 = \mu_2 = np(1 - p) = npq;$$

$k_3 = \mu_3 = np(1-p)(1-2p) = npq(q-p);$

$$k_4 = np(1-p)(1-6p+6p^2) = npq(1-6pq).$$

Hypergeometric Probability Distribution

There are many experiments in which the condition of independence is violated and the probability of success does not remain constant for all trials. Such experiments are called hypergeometric experiments. In other words, a hypergeometric experiments has the following properties.

- The outcomes of each trial may be classified into one of two categories, success and failure.
- The probability of success changes on each trial.
- The successive trials are dependent.
- The experiment is repeated a fixed number of times.

The number of successes, X in a hypergeometric experiment is called a hypergeometric r.v. and its probability distribution is called the hypergeometric distribution. When the hypergeometric r.v. X assumes a value x, the hypergeometric p.d. is given by the formula

$$P(X = x) = \frac{\binom{k}{x}\binom{N-k}{n-x}}{\binom{N}{n}},$$

$for\, x\, such that\, x = 0,1,2,\ldots,n\, and\, x = 0,1,2,\ldots,k.$

Where $N =$ number of units in the set or population, (a positive integer)

$n =$ number of units in the subset or sample (a positive integer) and

$k =$ number of successes in the set or population (a non negative integer less than or equal to N)

The hypergeometric p.d. has three parameters $N, n\, and\, k,$ or $N, n\, and\, p = \frac{k}{N})$, and is generally denoted by $h(x; N, n, k)$.

The hypergeometric p.d. is appropriate when

- A random sample of size n is drawn without replacement from a finite population of N units;
- k of the units are of one kind (classified as success) and the remaining $N-k$ of another kind (classifies as failure).

Applications for the hypergeometric distribution are found in many areas, with heavy use in acceptance sampling, electronic testing, and quality assurance. Obviously, in many of these fields, testing is done at the expense of the item being tested. That is, the item is destroyed and hence cannot be replaced in the sample. Thus, sampling without replacement is necessary. A simple example with playing cards will serve as our first illustration.

Properties of Hypergeometric Distribution

The important properties of the hypergeometric probability distribution are given here.

- The mean and variance of the hypergeometric probability distribution are $\mu = np$ and $\sigma^2 = npq\dfrac{N-n}{N-1}, where\ p = \dfrac{k}{N} and\ q = \dfrac{N-k}{N}$.
- If N becomes indefinitely large, the hypergeometric probability distribution tends to the binomial probability distribution.

CHAPTER 6

DISCRETE PROBABILITY DISTRIBUTION-II

INTRODUCTION

In previous chapter, we discussed some discrete probability distribution, their properties and application in real life problem. In this chapter we will discuss some more distributions.

Poisson Distribution

The Poisson distribution, named after the French mathematician Siméon Denis Poisson (1781-1840) who published its derivation in 1837, is used as

- A limiting approximation of the binomial distribution when the probability of success is very small but , the number of trials is so large that the product is of a moderate size;
- A distribution in its own right by considering a Poisson process where events occur randomly over a specified interval of time or space or length. Such random events might be the number of deaths by horse kicks per year; the number of telephone calls received per minute at some switch board; the number of taxicab arrivals at an intersection per day; the number of persons born blind per year in a large city; the number of typing errors per page in a book, the number of red blood cells in a specimen of blood; the number of radioactive particles emitted in a given period; the number of flaws per unit length of some

material, the number of insurance claims made to a company in a given time; etc.

Generally, most statisticians use Poisson approximation when p is $0.05\, or\, less$ and n is 20 or more, but in fact, the larger n is and the smaller p is, the better will be the approximation. If we assume that n goes to infinity and p approaches zero in such a way that $\mu = np$ remains constant, then the limiting form of the binomial probability distribution is

$$\lim_{\substack{n\to\infty \\ p\to 0}} b(x;n,p) = \frac{\mu^x.e^{-\mu}}{x!}, x = 0,1,2,\ldots,\infty$$

Where $e = 2.71828.$ The Poisson distribution has only one parameter $\mu > 0,$ and is denoted by $p(x,\mu).$ The parameter μ may be interpreted as the mean rate of occurrence of events. It is relevant to note that this is a probability distribution as the function is obviously non-negative, i.e. $p(x;\mu) \geq 0$ and it sums to unity, i.e.

$$\sum_{x=0}^{\infty} p(x;\mu) = \sum_{x=0}^{\infty} P(X = x)$$

$$= \sum_{x=0}^{\infty} \frac{\mu^x.e^{-\mu}}{x!} = e^{-\mu} \sum_{x=0}^{\infty} \frac{\mu^x}{x!}$$

$$= e^{-\mu}\left[1 + \mu + \frac{\mu^2}{2!} + \frac{\mu^3}{3!} + \ldots\right] = e^{-\mu}.e^{\mu} = 1$$

The Poisson probability distribution is also called the law of small numbers or the rare events distribution. It has found wide application in the field of biology, physics, operation research and management sciences. The Poisson distribution is appropriate when the number of possible occurrences is very large but the number of actual occurrences is very small in a fixed period of time.

Derivation of Poisson Approximation to the Binomial:

To derive an approximation formula to the binomial distribution $b(x;n,p)$ when $n \to \infty, p \to 0$ and the product np remains constant, we proceed as below:

The binomial distribution $b(x;n,p)$ may be written as

$$b(x;n,p) = \binom{n}{x} p^x q^{n-x}, for\; x = 0,1,\ldots,n.$$

$$= \frac{n(n-1)(n-2)\ldots(n-x+1)}{x!} p^x q^{n-x}$$

Let $np = \mu$. Then $p = \frac{\mu}{n}$ and $q = 1 - p = 1 - \frac{\mu}{n}$.

Replacing all terms involving p, we get

$$b(x;n,p) = \frac{n(n-1)(n-2)\ldots(n-x+1)}{x!}\left(\frac{\mu}{n}\right)^x \left(1 - \frac{\mu}{n}\right)^{n-x}$$

$$= \frac{\mu^x}{x!}\frac{n(n-1)(n-2)\ldots(n-x+1)}{x!}\left(1 - \frac{\mu}{n}\right)^n \left(1 - \frac{\mu}{n}\right)^{-x}$$

$$= \frac{\mu^x}{x!}\left[1.\left(1 - \frac{1}{n}\right)\left(1 - \frac{2}{n}\right)\ldots\left(1 - \frac{x-1}{n}\right)\right]\left(1 - \frac{\mu}{n}\right)^n \left(1 - \frac{\mu}{n}\right)^{-x}$$

Letting $n \to \infty\, and\; p \to 0$ so that $np = \mu$ remain constant, we observe that each of the terms $\left(1 - \frac{1}{n}\right), \left(1 - \frac{2}{n}\right), \ldots, \left(1 - \frac{x-1}{n}\right) and \left(1 - \frac{\mu}{n}\right)^{-x}$ approaches unity. The term $\left(1 - \frac{\mu}{n}\right)^n$ may be written as

$$\left(1 - \frac{\mu}{n}\right)^n = \left[\left(1 - \frac{\mu}{n}\right)^{\frac{n}{\mu}}\right]^{\mu}$$

Writing $\frac{n}{\mu} = k,$ the expression becomes $\left[\left(1 - \frac{1}{k}\right)^k\right]^{\mu}$

If n increases indefinitely, so does k. Therefore $\left(1-\frac{1}{k}\right)^k$ tends to e^{-1}, where $e = 2.71828.$ Therefore

$$\lim_{n\to\infty}\left(1-\frac{\mu}{n}\right)^n = e^{-\mu}$$

Thus, the limiting value of $P(X = x)$ is given by the expression

$$\lim_{n\to\infty} b(x;n,p) = \frac{\mu^x}{x!}.1.1\ldots1.e^{-\mu} = \frac{\mu^x e^{-\mu}}{x!}, for\, x = 0,1,\ldots,\infty$$

In other words, if X is a binomial r.v. such that

$$P(X = x) = \binom{n}{x} p^x q^{n-x}, then$$

$$\lim_{\substack{n\to\infty \\ np=\mu}} P(X = x) = \frac{\mu^x.e^{-\mu}}{x!}, x = 0,1,2,\ldots,\infty$$

It is denoted by $p(x;\mu).$ Hence a r.v. X having p.d. $p(x;\mu)$ is said to have a poisson distribution with parameter $\mu.$

Example: If X is a poisson random variable with parameter $\mu = 2,$ find the probabilities for $x = 0,1,2,3\, or\, more$.

Solution: Here the Poisson distribution is

$$p(x;2) = \frac{e^{-2}(2)^x}{2!}(x = 0,1,2,\ldots)$$

The desired probabilities for $x = 0,1,2,3\, or\, more$ are computed as below:

$$P(X = 0) = p(0;2) = e^{-2} = 0.135335$$

$$P(X = 1) = p(1;2) = \frac{e^{-2}.2}{1!} = 2(0.135335) = 0.27067$$

$$P(X=2)=p(2;2)=\frac{e^{-2}.(2)^2}{2!}=\frac{4}{2}(0.135335)=0.27067,$$

$$P(X\geq 3)=1-P(X<3)$$

$$=1-\left[P(X=0)+P(X=1)+P(X=2)\right]$$

$$=1-\left[0.135335+0.27067+0.27067\right]$$

$$=1-0.676675=0.323325$$

***Example*:** Two hundred passengers have made reservations for an airplane flight. If the probability that a passenger who has a reservation will not show up is $0.01,$ what is the probability that exactly three will not show up?

***Solution*:** Let us regard a "no show" as success. Then this is essentially a binomial experiment with $n=200$ and $p=0.01.$ Since p is very small and n is considerably large, we shall apply the Poisson distribution, using $\mu=(200)(0.01)=2$

Therefore, if X represents the number of successes (not showing up), we have

$$P(X=3)=p(3;2)=\frac{(2)^3.e^{-2}}{3!}$$

$$=\frac{(8)(0.1353)}{3\times 2\times 1}\quad\left(\therefore e^{-2}=\frac{1}{(2.71828)^2}=0.1353\right)$$

$$=0.1804$$

***Example*:** The probability that a man aged 50 years will die within a year is $0.01125.$ What is the probability of 12 such men at least 11 will reach their fifty-first birthday?

***Solution*:** Here $p=0.01125$ and $n=12.$ We compute the desired probability by means of Poisson distribution because the probability of death is very small.

Therefore, $\mu = np = 12.(0.01125) = 0.135$, and the Poisson distribution is

$$p(x;0.135) = \frac{e^{-0.135}(0.135)^x}{x!}$$

Now the probability that no person will die i.e. all the 12 persons will survive, is

$$p(0;0.135) = e^{-0.135}$$

$$= 1 - 0.135 + \frac{(0.135)^2}{2!} - \frac{(0.135)^3}{3!} + \ldots.$$

$$= 0.8737$$

And the probability that 1 person will die, i.e. 11 persons will survive is

$$p(1;0.135) = \frac{e^{-0.135}(0.135)}{1!} = 0.1179$$

Hence the probability that at least 11 persons will survive

$$= p(0;0.135) + p(1;0.135)$$

$$= 0.8737 + 0.1179 = 0.9916$$

Example: A sampling plan calls for taking 100 items from a very large lot which is one percent defective. $Let\,X$ be the number of defective found in the sample of 100. Construct a table of probabilities $P(X = x)$, for $x = 0,1,2,3,4,$ using first the binomial probability distribution and then the Poisson approximation.

Solution: Let p denote the probability that an item is defective. Then the binomial probabilities with $n = 100,\ p = 0.01\,and\,q = 0.99$ are

$$P(X = x) = \binom{100}{x}(0.01)^x(0.99)^{100-x}, x = 0,1,2,3\,and\,4.$$

These probabilities by means of the Poisson approximation, using $\mu = np = (100)(0.01) = 1$ are

$$P(X = x) = p(x;1) = \frac{1.e^{-1}}{x!}, \text{for } x = 0,1,2,3 \text{ and } 4.$$

The desired probabilities, on simplification are given in the following table:

x	$P(X = x)$	
	Binomial	Poisson
0	0.3660	0.3679
1	0.3697	0.3697
2	0.1849	0.1839
3	0.0610	0.0613
4	0.0149	0.0153

These results indicate that the approximation is very good.

Example: Ten is the average number of oil tankers arriving each day at a certain port. The facilities at the port can handle at most 15 tankers per day. What is the probability that on a given day tankers have to be turned away?

Solution: Let *X* be the number of tankers arriving each day.

$$P(X > 15) = 1 - P(X \le 15) = 1 - \sum_{x=0}^{15} p(x;10)$$

$$= 1 - 0.9513 = 0.0487$$

Like the binomial distribution, the Poisson distribution is used for quality control, quality assurance, and acceptance sampling. In addition, certain important continuous distributions used in reliability theory and queuing theory depend on the Poisson process.

Poisson Frequency Distribution

When the Poisson distribution is multiplied by N, the number of sets or experiments, each of n trials, the resulting distribution is known as the Poisson frequency distribution, and is denoted by

$$f(x) = N.\frac{e^{-\mu}.\mu^{x}}{x!}, where\, x = 0,1,2,\ldots,\infty$$

Example: For a machine making parts, there is a small probability of 0.002 for a part to be defective. The parts are supplied in bundle of 10. Calculate approximately the number of bundles containing no defective, one defective or two defectives in a consignment of $10,000$ bundles, given that $e^{-0.02} = 0.9802$

Solution: Let p be the probability of a part being defective. Then $p = 0.002\, and\, n = 10.$ Since p is extremely small, we apply Poisson approximation, using $\mu = np = 10 \times 0.002 = 0.02$

Hence the approximate number of bundles containing no defective, one defective or two defectives are the terms for $x = 0,1,\text{and}\,2$

$$N.p(x;\mu) = 10,000 \times \frac{(0.02)^{x}\, e^{-0.02}}{x!}$$

Putting $x = 0,$ we get

$$10,000 \times e^{-0.02} = 10,000 \times 0.9802 = 9802$$

Putting $x = 1,$ we get

$$10,000 \times e^{-0.02} \times 0.02 = 10,000 \times 0.9802 \times 0.02 = 196$$

Putting $x = 2,$ we get

$$10,000 \times e^{-0.02} \times \frac{(0.02)^{2}}{2!} = 10,000 \times 0.9802 \times \frac{(0.02)^{2}}{2} = 2\, approx.$$

Example: In a certain industrial facility, accidents occur infrequently. It is known that the probability of an accident on any given day is 0.005 and accidents are independent of each other.

(a) What is the probability that in any given period of 400 days there will be an accident on one day?

(b) What is the probability that there are at most three days with an accident?

***Solution*:** Let X be a binomial random variable with $n=400$ and $p=0.005$. Thus,

$np = 400 \times 0.005 = 2$

Using the Poisson approximation,

$P(X=1) = e^{-2} \times 2^{1} = 0.271$ and

$$P(X \leq 3) = \sum_{x=0}^{3} e^{-2} . \frac{2^{x}}{x!} = 0.857$$

***Example*:** In a manufacturing process where glass products are made, defects or bubbles occur, occasionally rendering the piece undesirable for marketing. It is known that, on average, 1 in every 1000 of these items produced has one or more bubbles. What is the probability that a random sample of 8000 will yield fewer than 7 items possessing bubbles?

***Solution*:** This is essentially a binomial experiment with n = 8000 and p = 0.001. Since p is very close to 0 and *n* is quite large, we shall approximate with the Poisson distribution using

$\mu = (8000)(0.001) = 8$

Hence, if X represents the number of bubbles, we have

$$P(X < 7) = \sum_{x=0}^{6} b(x; 8000, 0.001) \approx p(x; 8) = 0.3134$$

Properties of the Poisson Distribution

Some of the main properties of the Poisson distribution are given below:

- If the random variable X has a Poisson distribution with parameter μ, then its mean and variance are given by $E(X) = \mu \, and \, Var(X) = \mu$.

By definition,

$Mean = E(X)$

$$= \sum_{x=0}^{\infty} x . p(x; \mu), where \, p(x; \mu) = \frac{e^{-\mu} . \mu^{x}}{x!}$$

$$= 0.e^{-\mu} + 1.\mu e^{-\mu} + 2.\frac{\mu^2}{2!}e^{-\mu} + 3.\frac{\mu^3}{3!}e^{-\mu} + \ldots$$

$$= \mu\ .e^{-\mu}\left[1 + \mu + \frac{\mu^2}{2!} + \frac{\mu^3}{3!} + \ldots\right]$$

$$= \mu.e^{-\mu}.e^{\mu} = \mu$$

We can prove it by an alternative method

$$E(X) = \sum_{x=0}^{\infty} x\ .\frac{e^{-\mu}.\mu^x}{x!}$$

$$= e^{-\mu}\sum_{x=0}^{\infty}\frac{x.\mu^x}{x(x-1)!}$$

$$= \mu.e^{-\mu}\sum_{x=1}^{\infty}\frac{\mu^{x-1}}{(x-1)!}$$ (Since the first term in the summation being zero is omitted)

Let $y = x-1,$ then

$$E(X) = \mu.e^{-\mu}\sum_{y=0}^{\infty}\frac{\mu^y}{y!} \qquad (y = 0,1,2,\ldots\infty)$$

$$= \mu.e^{-\mu}.e^{\mu} = \mu$$

Hence mean is $\mu,$ the parameter of the distribution.

Now, $Var(X) = E(X^2) - [E(X)]^2$,where

$$E(X^2) = E[X(X-1) + X] = E(X) + E[X(X-1)]$$

$$= \sum_{x=0}^{\infty} x.\frac{e^{\mu}.\mu^x}{x!} + \sum_{x=0}^{\infty} x(x-1).e^{-\mu}\ .\frac{\mu^x}{x!}$$

$$= \mu + e^{-\mu}\sum_{x=0}^{\infty} x(x-1).\frac{\mu^x}{x(x-1)(x-2)!}$$

$$= \mu + \mu^2 . e^{-\mu} \sum_{x=2}^{\infty} \frac{\mu^{x-2}}{(x-2)!}$$ (x starts at 2, as the first two terms in the summation are zero)

Let $y = x - 2,$ then

$$E(X^2) = \mu + \mu^2 . e^{-\mu} \sum_{y=0}^{\infty} \frac{\mu^y}{y!} \quad (y = 0,1,2,\ldots)$$

Hence $Var(X) = \mu + \mu^2 - \mu^2 = \mu$

We observe that the Poisson distribution has an interesting property that its mean equals its variance.

- Higher moments of the distribution are found as below:

By definition, $\mu_3' = E(X^3) = \sum_{x=0}^{\infty} x^3 . p(x; \mu)$

Writing x^3 in the factorial form, i.e.

$x^3 = x(x-1)(x-2) + 3x(x-1) + x$

We have

$$\mu_3' = e^{-\mu} \sum_{x=0}^{\infty} [x(x-1)(x-2) + 3x(x-1) + x] . \frac{\mu^x}{x!}$$

$$= e^{-\mu} . \mu^3 \sum_{x=3}^{\infty} \frac{\mu^{x-3}}{(x-3)!} + 3\mu^2 . e^{-\mu} \sum_{x=2}^{\infty} \frac{\mu^{x-2}}{(x-2)!} + \mu . e^{-\mu} \sum_{x=1}^{\infty} \frac{\mu^{x-1}}{(x-1)!}$$

$$= e^{-\mu} . \mu^3 \sum_{y=0}^{\infty} \frac{\mu^y}{y!} + 3\mu^2 e^{-\mu} \sum_{y=0}^{\infty} \frac{\mu^y}{y!} + \mu . e^{-\mu} \sum_{y=0}^{\infty} \frac{\mu^y}{y!}$$

$$= \mu^3 + 3\mu^2 + \mu, and$$

$$\mu_4' = E(X^4) = \sum_{x=0}^{\infty} x^4 . p(x;, \mu)$$

writing x^4 in the factorial form as

$x^4 = x(x-1)(x-2)(x-3) + 6x(x-1)(x-2) + 7x(x-1) + x$, we get

$$\mu_4^{'} = e^{-\mu}.\sum_{x=4}^{\infty}\frac{\mu^{x-4}}{(x-4)!}+6\mu^3 e^{-\mu}.\sum_{x=3}^{\infty}\frac{\mu^{x-3}}{(x-3)!}+7\mu^2.e^{-\mu}.\sum_{x=2}^{\infty}\frac{\mu^{x-2}}{(x-2)!}+\mu.e^{-\mu}.\sum_{x=1}^{\infty}\frac{\mu^{x-1}}{(x-1)!}$$

$$= \mu^4 + 6\mu^3 + 7\mu^2 + \mu$$

Hence $\mu_3 = \mu_3^{'} - 3\mu_1^{'}.\mu_2^{'} + 2(\mu_1^{'})^3$

$$= \mu^3 + 3\mu^2 + \mu) - 3\mu(\mu^2 + \mu) + 2\mu^3 = \mu;$$

$$\mu_4 = \mu_4^{'} - 4\mu_1^{'}\mu_3^{'} + 6\mu_2^{'}(\mu_1^{'})^2 - 3(\mu_1^{'})^4,$$

$$= (\mu^4 + 6\mu^3 + 7\mu^2 + \mu) - 4\mu(\mu^3 + 3\mu^2 + \mu) + 6\mu^2(\mu^2 + \mu) - 3\mu^4$$

$$= 3\mu^2 + \mu.$$

Now $\beta_1 = \frac{\mu_{3^2}}{\mu_2^3} = \frac{\mu^2}{\mu^3} = \frac{1}{\mu}$, *and*

$$\beta_2 = \frac{\mu_4}{\mu_2^2} = \frac{3\mu^2 + \mu}{\mu^2} = 3 + \frac{1}{\mu}.$$

- The shape of the Poisson distribution depends on the value of its parameter μ. As the distribution takes on an infinite number of x values (theoretically), the distribution will be positively skewed. The distribution tends to be symmetrical as μ becomes larger and larger.
- Reproductive Property. If two independent r.v. 's X *and* Y have Poisson distribution with parameters μ and v, then their sum $X+Y$ has also a Poisson distribution with parameter $\mu+v$.

Proof: Here X is $p(x;\mu)\, and\, Y\, is\, p(y;v);$ and we desire to find $(X+Y=k)\ for\ k=0,1,2,\ldots.$

For $k=0, P(X+Y=0) = P(X=0)P(Y=0)(\therefore X\ and\ Y\ are\ independent)$

For $k=1, P(X+Y=1) = P(X=0).P(Y=1) + P(X=1).P(Y=0)$

$$= e^{-\mu}.ve^{-v} + \mu.e^{-\mu}.e^{-v}$$

$= e^{-(\mu+v)}(\mu+v)$

Similarly for $k=2, P(X+Y=2)=\frac{e^{-(\mu+v)}}{2!}(\mu+v)^2.$

And, in general for $X+Y=k$ when $X = i \, and \, Y = k-i \; for \, i = 0,1,2,\ldots,k,$ we have

$$P(X+Y=k)=P(X=0)P(Y=k)+P(X=1).P(Y=k-1)+\ldots+P(X=k).P(Y=0)$$

$$=\frac{e^{-\mu}}{0!}.\frac{v^k e^{-v}}{k!}+\frac{\mu e^{-\mu}}{1!}.\frac{v^{k-1}e^{-v}}{(k-1)!}+\ldots+\frac{\mu^k e^{-\mu}}{k!}.\frac{e^{-v}}{0!}$$

$$=e^{-(\mu+v)}\left[\frac{v^k}{k!}+\mu\frac{v^{k-1}}{(k-1)!}+\ldots+\frac{\mu^k}{k!}\right]$$

Multiplying the r.h.s. by $\frac{k!}{k!},$ we get

$$P(X+Y=k)=\frac{e^{-(\mu+v)}}{k!}(\mu+v)^k,$$

Which is a Poisson distribution with parameter $\mu+v.$ This result can be easily generalized.

The converse of this result, that is, if $X \, and \, Y$ are independent, and $X+Y$ has a Poisson distribution, then each of the r.v. X and Y has a Poisson distribution, is also true and was proved by Raikov.

The Recurrence formula for the Poisson Distribution

Let X be a Poisson random variable with parameter $\mu.$ Then,

$$P(X=x)=e^{-\mu}.\frac{\mu^x}{x!}, and$$

$$P(X=x-1)=e^{-\mu}.\frac{\mu^{x-1}}{(x-1)!}$$

Dividing, we get

$$\frac{P(X=x)}{P(X=x-1)}=\frac{e^{-\mu}.\mu^{x}}{x!}\cdot\frac{(x-1)!}{e^{-\mu}.\mu^{x-1}}=\frac{\mu}{x}$$

Therefore $P(X=x)=\frac{\mu}{x}.P(X=x-1)\ for\, x=1,2,3,\ldots.$ is the recurrence formula for the Poisson distribution. Using this recurrence relationship, the Poisson probabilities can be obtained more easily.

Fitting a Poisson Distribution to Observed Data

The Poisson distribution can be fitted (or derived) if we know the value of its mean which is usually obtained equating to the mean estimated from the observed frequency distribution, provided that the probability of occurrence is very small. Using this vale of mean, the expected or theoretical frequencies are computed. The following classical example will illustrate the procedure.

Example: Bortkiewicz collected data on the number of deaths from horse kicks in Prussian Army Corps over a period of 20 years. The distribution of deaths was as follows

No.of Deaths	0	1	2	3	4	5	*Total*
Frequency	109	65	22	3	1	0	200

Fit a Poisson distribution to these data and compute the theoretical frequencies.

Solution: To fit a Poisson distribution, we need to compute (estimate) the value of mean of the given distribution and equate it to $\mu.$ the mean of the Poisson distribution.

Now $\bar{x}=\frac{\Sigma f_i x_i}{\Sigma f_i}=\frac{0+65+44+9+4}{200}$

$=\frac{122}{200}=0.61,$ which is an estimate of $\mu.$

Hence the fitted Poisson distribution is given by

$$P(X=x)=p(x;0.61)=\frac{e^{-0.61}.(0.61)^x}{x!}, where\, x=0,1,2,3,\ldots$$

The theoretical or expected frequencies of x deaths are computed by multiplying the probabilities by N, which is 200 here. The probabilities are computed by using the Poisson recurrence formula, which takes the form

$$P(X=x)=\frac{0.61}{x}.P(X=x-1)\; for\, x=1,2,3,4,5$$

In case the table values for $e^{-\mu}$ are not available, they are computed by use of logarithms as below:

No.of deaths (x)	*Probability*	*Expected Frequency*
0	$e^{-0.61}=0.5434$	108.68
1	$e^{-0.61}\frac{(0.61)}{1}=0.3315$	66.30
2	$e^{-0.61}\frac{(0.61)^2}{2!}=0.1011$	20.22
3	$e^{-0.61}\frac{(0.61)^3}{3!}=0.0206$	4.12
4	$e^{-0.61}\frac{(0.61)^4}{4!}=0.0031$	0.62
5		
		200.002

The Poisson Process

In an earlier section, the Poisson distribution was obtained as a limiting approximation to the binomial distribution. The terms of the Poisson distribution can also be derived from a Poisson process which may be defined as a physical process governed at least in part by some random mechanism. The occurrence of traffic deaths per month in a city is an example of a Poisson process. A Poisson process has the following properties.

- The probability that an event occurs in a very short time interval Δ, is proportional to the length of the time interval, i.e. is approximately λ Δ, when λ is a positive quantity and may be interpreted as the average number of occurrences per unit of time.
- The probability that two or more events occur in such a short interval is so small that it can be neglected.
- Events occurring that two or more events occur in such a short interval is so small that it can be neglected.
- Events occurring in non- overlapping intervals of time are statistically independent.

These properties are assumed to hold for events occurring randomly in regions of space, etc. Under these properties, it can be shown that the probability for the number of occurrences of a random event in an interval of stated length t is given by the Poisson distribution with the parameter λt. That is, the Poisson process formula is

$$p(x;\lambda t)=\frac{e^{-\lambda t}(\lambda t)^x}{x!},$$

where t = number of units of time,

x= number of occurrences in t units of time, and

λ = average number of occurrences per unit of time.

Example: Telephone calls are being placed through a certain exchange at random times on the average of four per minute. Assuming Poisson process, determine the probability that in a 15- second interval, there are 3 or more calls.

Solution: Taking a minute as the unit of time, we have

$\lambda = 4\,calls\ per\ minute\ = 4\,calls\ per\ 60\,seconds$

As 15 seconds is $\frac{15}{60}=\frac{1}{4}$ units of time, so $t=\frac{1}{4}$ and therefore the average number of calls per 15 seconds interval i.e. $\lambda t = 4\times\frac{1}{4}=1.$

Hence, using the Poisson process formula $p(x\,;\lambda t)=\frac{e^{-\lambda t}(\lambda t)^x}{x!}$,

we have

$$P(3\,or\,more\,calls\,in\,15-second\,interval)$$

$$=1-P(0,1\,or\,2\,calls\,in\,15-second\,interval)$$

$$=1-\sum_{x=0}^{2}p(x\,;\lambda t)$$

$$=1-\sum_{x=0}^{2}\frac{e^{-1}(1)^x}{x!}$$

$$=1-\sum_{x=0}^{2}\frac{(0.3679)(1)^x}{x!}$$

$$=1-0.91975$$

$$=0.08025$$

***Example*:** Flaws in a certain type of drapery material appear on the average of one in 150 square feet. If we assume the Poisson distribution, find the probability of at most one flaw in 225 square feet.

***Solution*:** Taking 150 square feet as the unit of area, we have

$$\lambda=1\,flaw\,per\,150\,square\,feet.$$

As 225 square feet are $\frac{225}{150}=1.5$ units of area, so $t=1.5$ and therefore the average number of flaws per 225 square feet, i.e. $\lambda t=1\times1.5=1.5.$ Assuming the flaws are a Poisson process, we have

$$P(at\,most\,one\,flaw\,in\,225\,square\,feet)=\sum_{x=0}^{1}p(x\,;\lambda t)$$

$$=\sum_{x=0}^{1}\frac{e^{-1.5}(1.5)^{x}}{x!},\quad \text{where } e^{-1.5}=0.2231$$

$$=0.2231+0.3347=0.5578$$

Moment Generating and Cumulant Generating Functions of the Poisson Distribution

The $m.g.f$ for the Poisson distribution with respect to the origin, is found as below:

$$M_0(t)=E\left(e^{tx}\right)=\sum_{x=0}^{\infty}e^{tx}\,p(x\ ;\mu)$$

$$=\sum_{x=0}^{\infty}e^{tx}.\frac{\mu^{x}e^{-\mu}}{x!}=e^{-\mu}\sum_{x=0}^{\infty}\frac{\left(\mu e^{t}\right)^{x}}{x!}$$

$$=e^{-\mu}\ .e^{\mu\,e^{t}}$$

$$=e^{\mu\left(e^{t}-1\right)}$$

The mean and variance are obtained as

$$\mu_1^{'}=E(X)=\left[\frac{d}{dt}\left(e^{\mu\left(e^{t}-1\right)}\right)\right]_{t=0}$$

$$=\left[\mu e^{t}\left(e^{\mu\left(e^{t}-1\right)}\right)\right]_{t=0}=\mu$$

And $$\mu_2^{'}=E\left(X^2\right)=\left[\frac{d^2}{dt^2}\left(e^{\mu\left(e^{t}-1\right)}\right)\right]_{t=0}$$

$$=\left[\mu\ .e^{t}.\left(e^{\mu\left(e^{t}-1\right)}\right)+\mu^{2}\ .\ e^{2t}.\left(e^{\mu\left(e^{t}-1\right)}\right)\right]_{t=0}$$

$$=\mu+\mu^{2}$$

Hence the variance is $\sigma^{2}=\mu_2^{'}-\mu_1^{'2}=\mu+\mu^{2}-\mu^{2}=\mu$

The cumulant generating function *(c. g . f)* of the Poisson distribution with respect to origin is given by

$$k(t) = \log_e M_0(t)$$

$$= \mu\left(e^t - 1\right)$$

$$= \mu\left(t + \frac{t^2}{2!} + \frac{t^3}{3!} + \ldots + \frac{t^r}{r!} + \ldots\right)$$

$$k_r = co - efficient\, of\, \frac{t^r}{r!} in\, k(t) = \mu, for\, all\, r$$

Hence all the cumulant of the Poisson distribution are equal to μ.

Example: Prove the following recurrence formula for a Poisson distribution $p(x;m)$:

$$\mu_{r+1} = r m \mu_{r-1} + m\frac{d\mu_r}{dm}$$

By definition

$$\mu_r = \sum_{x=0}^{\infty} p(x;m)(x-m)^r$$

$$= \sum_{x=0}^{\infty} \frac{e^{-m} m^x}{x!}(x-m)^r$$

Differentiating with respect to m, we get

$$\frac{d\mu_r}{dm} = \sum_{x=0}^{\infty}\left[-r.\frac{e^{-m} m^x}{x!}(x-m)^{r-1} + x.\frac{e^{-m} m^{x-1}}{x!}(x-m)^r - \frac{e^{-m} m^x}{x!}(x-m)^r\right]$$

Multiplying both sides by m and simplifying, we get

$$m\frac{d\mu_r}{dm} = r_0\mu + \sum_{x=0}^{\infty}\frac{e^{-m}.m^x}{x!}(x-m)^r.(x-m)$$

$$= -rm\,\mu_{r-1} + \mu_{r+1}$$

Hence
$$\mu_{r+1} = rm\,\mu_{r-1} + m\frac{d\mu_r}{dm}$$

Negative Binomial Distribution

In the binomial experiments, the number of successes varies and the number of trials is fixed. But there are experiments in which the number of successes is fixed and the number of trials varies to produce the fixed number of successes. Such experiments are called negative binomial experiments. In other words, a negative binomial experiment possesses the following properties:

- The outcomes of each trial may be classified into one of two categories: success (S) and failure (F).
- The probability of success, denoted by p, remains constant for all trials.
- The successive trials are all independent.
- The experiment is repeated a variable number of times to obtain a fixed number of successes.

When X denotes the number of trials to produce k successes in a negative binomial experiment, it is called a negative binomial variable, and its p.d., is called the negative binomial distribution. When the negative binomial r.v X assumes a value x on which the kth success occurs, the negative binomial distribution is given by

$$P(X=x)=\binom{x-1}{k-1}p^k q^{x-k}, \quad x=k,k+1,k+2,\ldots$$

The negative binomial p.d. has two parameter k and $p>0,$ and is generally denoted by $b(x;k,p).$ To ensure that, for the negative binomial distribution, the probability function sums to one, we proceed as below:

$$\Sigma b(x;k,p)=\sum_{x=k}^{\infty}\binom{x-1}{k-1}p^k q^{x-k}, (x=k,k+1,k+2,\ldots)$$

Let $y=x-k,$ then

$$=\sum_{y=0}^{\infty}\binom{y+k-1}{k-1}p^k q^y, (y=0,1,2,\ldots)$$

$$= p^k \left[\binom{k-1}{k-1} q^0 + \binom{k}{k-1} q + \binom{k+1}{k-1} q^2 + \ldots \right]$$

$$= p^k \left[1 + kq + \frac{k(k+1)}{2!} q^2 + \ldots \right]$$

$$= p^k [1-q]^{-k} = p^k . p^{-k} = 1$$

The distribution takes its name from the fact that $\binom{x-1}{k-1} p^k q^{x-k}$ is a term in the expansion of $p^k (1-q)^{-k}$, a binomial with negative index. Thus the probabilities at $kth,\ (k+1)th, (k+2)th, \ldots$ trial are $p^k \left[1, kq, \frac{k(k+1)}{2} q^2, \ldots \right]$

This distribution is sometimes also called the Pascal distribution, after the French mathematician Blaise Pascal (1623-1662). The distribution is found to occur in many biological situations and inverse sampling from a binomial population.

Derivation of the Negative Binomial Distribution

The negative binomial distribution can be derived in various ways. The following derivation is based on the Bernoulli trials.

To find an expression for the probability that x trials are made to achieve k successes with the condition that the last trial must be a success, we proceed as follows:

A sequence containing k successes in exactly x independent trials with the condition that it ends in a success, can be obtained as

$$\underbrace{SS \ldots S}_{k-1 \ times} \quad \underbrace{FF \ldots FS}_{x-k \ times}$$

Thus the probability of a success on the xth trial preceded by $(k-1)$ successes and $(x-k)$ failures, is $p^{k-1} q^{x-k} p = p^k q^{x-k}$

Since the last trial must be a success, therefore the total number of mutually exclusive ways in which $(k-1)$ successes and $(x-k)$ failures preceding the last success can occur in any order, is given by

$$\binom{x-1}{k-1}$$

Hence the formula for the probability that kth success occurs on the xth trial is

$$P(X=x)=\binom{x-1}{k-1}p^k q^{x-k},\ where\, x=k,k+1,k+2,\ldots$$

The negative binomial distribution can also be obtained when two or more Poisson r.v. are combined term by term.

Example: A person throws a pair of fair dice. What is the probability that he will get a total of 7 for the second time on the eighth throw?

Solution: The probability that he gets a total of $7\,is\,\frac{6}{36}, i.e. p=\frac{1}{6}.$

Since the number of successes is fixed, therefore the negative binomial distribution with $k=2$ (second success) and $x=8$ is used.

Hence $$P(X=8)=\binom{8-1}{2-1}\left(\frac{1}{6}\right)^2\left(\frac{5}{6}\right)^{8-2}=0.0651$$

Example: Three people each toss a coin and the odd man pays for the coffee. If the coins all show heads or all show tails, they are tossed again. What is the probability that a decision is reached in five tosses or fewer?

Solution: To reach a decision on any trial, the coins must result in either "2 heads and 1 tail" or "2 tails and 1 head". The probability of these events is computed by the binomial distribution because

- Each coin has two possible results; a head or a tail.
- The probability of getting a head is $p=\frac{1}{2}$ and remains the same for each coin.
- Three people toss the coin independently.

- Three coins are tossed in each case. (i.e. $n = 3$)

Therefore

$$P(2\,heads\,and\,1\,tail) = \binom{3}{2}\left(\frac{1}{2}\right)^2\left(\frac{1}{2}\right) = \frac{3}{8}$$

$$P(1\,head\,and\,2\,tails) = \binom{3}{1}\left(\frac{1}{2}\right)\left(\frac{1}{2}\right)^2 = \frac{3}{8}$$

Hence the probability of reaching a decision is $\frac{3}{8} + \frac{3}{8} = \frac{6}{8} = 0.75$

Next, we find the probability that a decision is reached in 5 tosses or fewer. We observe that

- We reach or do not reach a decision with each set
- The probability of reaching a decision is $p = 0.75$ and remains the same for all sets.
- The results of all sets are independent,
- A variable number of sets is required to produce 1 decision.

We therefore compute the required probability by means of the negative binomial distribution, where

$k = 1, p = 0.75\, and\, x \le 5(the\, number\, of\, trials)$.

Thus,

$$\sum_{x=1}^{5}\binom{x-1}{k-1}(0.75)^k(0.25)^{x-k}, k = 1$$

$$= 0.75 + 0.1875 + 0.0469 + 0.0117 + 0.0029$$

$$= 0.9990$$

Properties of the Negative Binomial Distribution

The important properties of the negative binomial distribution are given below:

- The mean of the negative binomial distribution is less than its variance.

We find the mean and variance by deriving the m.g.f of the negative binomial distribution.

The m.g.f about the origin is

$$M_0(t) = E\left(e^{tX}\right)$$

$$= p^k \sum_{x=0}^{\infty} \binom{x+k-1}{k-1} q^x . e^{tx}$$

$$= p^k \sum_{x=0}^{\infty} \binom{x+k-1}{k-1} \left(qe^t\right)^x$$

$$= p^k \left(1-qe^t\right)^{-k}$$

Now, $$Mean = E(X) = \left[\frac{dM_0(t)}{dt}\right]_{t=0}$$

$$= \left[p^k .kqe^t\left(1-qe^t\right)^{-k-1}\right]_{t=0}$$

$$= p^k .kq(1-q)^{-k-1}$$

$$= kq\, p^k p^{-k-1} = \frac{kq}{p}$$

And $\sigma^2 = E\left(X^2\right) - \left[E(X)\right]^2$, where

$$E\left(X^2\right) = \left[\frac{d^2 M_0(t)}{dt^2}\right]_{t=0}$$

$$=[p^k .kqe^t\left(1-qe^t\right)^{-k-1} + p^k .k(k+1)q^2 p^k (1-q)^{-k-2}$$

$$= \frac{kq}{p} + \frac{k(k+1)q^2}{p^2}$$

Thus $$\sigma^2 = \frac{kq}{p} + \frac{k(k+1)}{p^2} - \left(\frac{kq}{p}\right)^2$$

$$=\frac{kqp+k^2q^2+kq^2-k^2q^2}{p^2}$$

$$=\frac{kq(p+q)}{p^2}$$

$$=\frac{kq}{p^2}$$

The variance will be greater than mean, if

$$\frac{kq}{p^2}>\frac{kq}{p} \; or \; if \; \frac{1}{p^2}>\frac{1}{p}$$

i.e. if $1>p$, which is obviously true.

Hence we observe that the variance of the negative binomial distribution is greater than its mean. This is an important feature of this distribution.

- The negative binomial distribution is always positively skewed.

Geometric Distribution

When an experiment consists of independent trials with probability p of success and the trials are repeated until the first success occurs it is called a geometric experiment. In other words, a geometric experiment has the following four properties:

- The outcomes of each trial may be classified into one of two categories, success and failure.
- The probability of success p remains constant for all trials.
- The successive trials are all independent.
- The experiment is repeated a variable number of times until the first success is obtained.

If X represents the number of trials needed for the first success, then X is called a geometric r.v. and its p.d. is called the geometric distribution.

It has only one parameter p and is denoted by $g(x\,;p)$. The geometrical distribution derives its name from the fact that its successive terms constitute a geometric progression. Since a geometric r.v. represents how long one has to wait for a success, it is also called a waiting time r.v. It is interesting to note that a geometric distribution is a special case of a negative binomial distribution when $k=1$.

Derivation of the Geometric Distribution

Let the random variable X denote the number of trials required up to and including the first success of an event. Then X takes the values $1,2,3,\ldots,\infty$. Since $X=x$ if and only if the first $(x-1)$ trials result in failures and, the xth trial yields a success in the $Bernoullian$ sequence of trials, we therefore have the probability distribution of X, as

$P(X=x)=q^{x-1}p, x=1,2,3,\ldots,\infty.$

This is obviously a probability distribution as

- $P(X=x)\geq 0$
- The sum of the probabilities is unity, i.e.

$$\sum_{x=1}^{\infty}P(X=x)=p+qp+q^2p+q^3p+\ldots$$

$$=p\left[1+q+q^2+q^3+\ldots\right]$$

$$=p[1-q]^{-1}=pp^{-1}=1$$

Example: If the probability that a person will believe a rumor about the retirement of a certain politician is 0.25, what is the probability that

- The sixth person to hear the rumor will be the first to believe it;
- The twelfth person to hear the rumor will be the fourth to believe it?

Solution: Let X denote the number of a person who hears the rumor. Then the number of a person who believes it, will be considered a success.

- Since the sixth person is the first to believe the rumor, i.e. the first success occurs on the sixth trial, therefore the geometric distribution with $p=0.25\, and\, x=6$ is appropriate.

Hence, using the geometric distribution $P(X=x)=pq^{x-1}$, we get

$P(X=6)=(0.25)(0.75)^5=0.059.$

- Since the twelfth person hearing the rumour will be the fourth to believe implies that the 4th success occurs on the 12th trial, therefore the negative binomial distribution with $p=0.25, k=4\, and\, x=12$ is appropriate.

Hence, using the negative binomial distribution

$$b^*(x;k,p)=\binom{x-1}{k-1}p^k q^{x-k}, we\, get$$

$$b^*(12;4,0.25)=\binom{12-1}{4-1}(0.25)^4(0.75)^{12-4}$$

$$=\binom{11}{3}\left(\frac{1}{4}\right)^4\left(\frac{3}{4}\right)^8$$

$$=\frac{165\times 6561}{265\times 65536}$$

$$=0.0645$$

Properties of a Geometric Distribution

- The mean and variance of the geometric distribution are $\mu=\frac{1}{p}$ and $\sigma^2=\frac{q}{p^2}$

Let the r.v. X have a geometric distribution $g(x;p)=pq^{x-1}$. Then

$$\mu=E(X)=\Sigma x.g(x;p)$$

$$=\sum_{x=1}^{\infty} x.q^{x-1}p,\ where\, x=1,2,3,\ldots,\infty.$$

$$= p + 2qp + 3q^2 p + 4q^3 p + \ldots$$

$$= p\left[1 + 2q + 3q^2 + 4q^3 + \ldots\right]$$

$$= p[1-q]^{-2}$$

$$= p.p^{-2} = \frac{1}{p}$$ and

$$\sigma^2 = E\left(X^2\right) - \left[E\left(X\right)\right]^2$$, where

$$E\left(X^2\right) = \sum_{x=1}^{\infty} x^2 . q^{x-1} p$$

$$= p + 2^2 qp + 3^2 q^2 p + 4^2 q^3 p + \ldots$$

$$= p\left[1 + 4q + 9q^2 + 16q^3 + \ldots\right]$$

$$= p\left[\left(1 + 3q + 6q^2 + 10q^3 + \ldots\right) + \left(q + 3q^2 + 6q^3 + \ldots\right)\right]$$

$$= p\left[(1-q)^{-3} + q(1-q)^{-3}\right]$$

$$= \frac{1}{p^2} + \frac{q}{p^2} - \left(\frac{1}{p}\right)^2 = \frac{q}{p^2}.$$

- The distribution is positively skewed.

Moment Generating Function of the Geometric Distribution

The m.g.f. of the geometric distribution is derived as below:

By definition, $M_0(t) = E\left(e^{tx}\right) = \sum_{x=1}^{\infty} e^{tx} q^{x-1} p$

$$= pe^t \sum_{x=1}^{\infty} \left(e^t q\right)^{x-1}$$

$$= pe^t \left[1 + qe^t + \left(qe^t\right)^2 + \ldots\right]$$

$$= pe^t\left[1-qe^t\right]^{-1}$$

$$= \frac{pe^t}{1-qe^t}, where\, qe^t < 1$$

In order to differentiate the m.g.f., we write it as

$$M_0(t) = \frac{p}{e^{-t}-q} = p\left(e^{-t}-q\right)^{-1}$$

Thus $M_0'(t) = pe^{-t}\left(e^{-t}-q\right)^{-2}, and$

$$M_0''(t) = 2\,pe^{-2t}\left(e^{-t}-q\right)^{-3} - pe^{-t}\left(e^{-t}-q\right)^{-2}$$

Hence $E(X) = p(1-q)^{-2} = \frac{1}{p},$

$$E\left(X^2\right) = 2p(1-q)^{-3} - p(1-q)^{-2}$$

$$= \frac{2}{p^2} - \frac{1}{p};$$ and

$$\sigma^2 = E\left(X^2\right) - \left[E(X)\right]^2$$

$$= \frac{2}{p^2} - \frac{1}{p} - \left(\frac{1}{p}\right)^2 = \frac{q}{p^2}.$$

Multinomial Distribution

A binomial experiment becomes a multinomial experiment when there are more than two possible outcomes of each trial. For example, manufactured items may be classified as good, average, or inferior; or a road accident may result in no injury, minor injuries, severe injuries, or fatal injuries.

A multinomial experiment has the following properties:

- The outcomes of each trial may be classified into one of k mutually exclusive categories $C_1, C_2, \ldots, C_k$.

- The probability of the ith outcome is p_i which remains constant and $\sum p_i = 1$.
- The successive trials are all independent.
- The experiment is repeated a fixed number of times, say n.

Derivation of the Multinomial Distribution

If n independent trials be made, then one specified order in which C_1 occurs x_1 times, C_2 occurs x_2 times,…, C_k occurs x_k times, where $x_1 + x_2 + \ldots + x_k = n,$ is

$$\underbrace{C_1 \ldots C_1}_{x_1 times}\ \underbrace{C_2 \ldots C_2}_{x_2 times} \ldots\ldots\ldots \underbrace{C_k \ldots C_k}_{x_k times}$$

The probability of this happening by multiplicative law is

$$\underbrace{p_1 \cdots p_1}_{x_1 times}\ \underbrace{p_2 \cdots p_2}_{x_2 times} \ldots\ldots\ldots \underbrace{p_k \cdots p_k}_{x_k times} = p_1^{x_1}\, p_2^{x_2} \cdots p_k^{x_k}$$

But we are interested in events occurring in any order. Therefore, the total number of mutually exclusive orders in which this can happen, is

$$\binom{n}{x_1, x_2, \ldots, x_k} = \frac{n!}{x_1!x_2!,\ldots,x_k!}$$

Hence the required multinomial probability is

$$P(X_1 = x, X_2 = x_2, \ldots, X_k = x_k)$$

$$= \frac{n!}{x_1!x_2!,\ldots,x_k!}(p_1)^{x_1}\,(p_2)^{x_2} \ldots (p_k)^{x_k}$$

$$= \frac{n!}{\prod_{i=1}^{k} x_i!} \prod_{i=1}^{k} (p_i)^{x_i}$$, where $$\sum_{i=1}^{k} x_i = n \ \& \sum_{i=1}^{k} p_i = 1$$

The multinomial distribution takes its name from the fact that that the above probabilities correspond to the terms in the multinomial expansion of $(p_1 + p_2 + \ldots + p_k)^n$. The parameter of this distribution are

$n, p_1, p_2, \ldots, p_k$.

The mean and variance of the multinomial distribution are

$E(X_i) = np_i$ and $Var(X_i) = np_i q_i$

When $k = 2,$ the multinomial distribution reduces to the binomial probability distribution. Thus, binomial distribution is a special case of the multinomial distribution.

Example: A box contain 5 red, 4 white and 3 blue marbles. A sample of six marbles is drawn with replacement, i.e., each marble is replaced before the next one is drawn. Find the probability that out of 6 marbles selected, 3 are red, 2 are white and one is blue.

Solution: Let X_1, X_2 *and* X_3 denote the red, white and blue marbles. Then

$$p_1 = P(X_1 = 3) = \frac{5}{12}$$

$$p_2 = P(X_2 = 2) = \frac{4}{12}$$

$$p_3 = P(X_3 = 1) = \frac{3}{12}$$

Hence

$$P(X_1 = 3, X_2 = 2, X_3 = 1) = \frac{6!}{3!2!1!}\left(\frac{5}{12}\right)^3\left(\frac{4}{12}\right)^2\left(\frac{3}{12}\right)^1 = \frac{625}{5184}$$

SUMMARY AND DISCUSSION

In this chapter, we explored some fundamental concepts of discrete probability distribution. We also discussed their properties and applications as well.

Poisson Distribution

The Poisson distribution, named after the French mathematician Siméon Denis Poisson (1781-1840) who published its derivation in 1837, is used as

- A limiting approximation of the binomial distribution $b(x;n,p)$ when p, the probability of success is very small but n, the number of trials is so large that the product $np = \mu$ is of a moderate size;
- A distribution in its own right by considering a Poisson process where events occur randomly over a specified interval of time or space or length. Such random events might be the number of deaths by horse kicks per year; the number of telephone calls received per minute at some switch board; the number of taxicab arrivals at an intersection per day; the number of persons born blind per year in a large city; the number of typing errors per page in a book, the number of red blood cells in a specimen of blood; the number of radioactive particles emitted in a given period; the number of flaws per unit length of some material, the number of insurance claims made to a company in a given time; etc.

Generally, most statistician use Poisson approximation when p is $0.05\, or\, less$ and n is 20 or more, but in fact, the larger n is and the smaller p is, the better will be the approximation. If we assume that n goes to infinity and p approaches zero in such a way that $\mu = np$ remains constant, then the limiting form of the binomial probability distribution is

$$\lim_{\substack{n\to\infty \\ p\to 0}} b(x;n,p) = \frac{\mu^x . e^{-\mu}}{x!}, x = 0,1,2,\ldots,\infty$$

Where $e = 2.71828$. The Poisson distribution has only one parameter $\mu > 0$, and is denoted by $p(x,\mu)$. The parameter μ may be interpreted as the mean rate of occurrence of events. It is relevant to note that this is a probability distribution as the function is obviously non-negative, i.e. $p(x;\mu) \geq 0$ and it sums to unity, i.e.

$$\sum_{x=0}^{\infty} p(x;\mu) = \sum_{x=0}^{\infty} P(X = x)$$

$$= \sum_{x=0}^{\infty} \frac{\mu^x . e^{-\mu}}{x!} = e^{-\mu} \sum_{x=0}^{\infty} \frac{\mu^x}{x!}$$

$$= e^{-\mu}\left[1+\mu+\frac{\mu^2}{2!}+\frac{\mu^3}{3!}+\ldots\right] = e^{-\mu}.e^{\mu} = 1$$

The Poisson probability distribution is also called the law of small numbers or the rare events distribution. It has found wide application in the field of biology, physics, and operation research and management sciences. The Poisson distribution is appropriate when the number of possible occurrences is very large but the number of actual occurrences is very small in a fixed period of time.

Poisson Frequency Distribution

When the Poisson distribution is multiplied by $N,$ the number of sets or experiments, each of n trials, the resulting distribution is known as the Poisson frequency distribution, and is denoted by

$$f(x) = N.\frac{e^{-\mu}.\mu^x}{x!}, where\, x = 0,1,2,\ldots,\infty$$

Properties of the Poisson Distribution

Some of the main properties of the Poisson distribution are given below:

- If the random variable X has a Poisson distribution with parameter $\mu,$ then its mean and variance are given by $E(X) = \mu\, and\, Var(X) = \mu.$
- Higher moments of the distribution are found as below:

$$\mu_3' = \mu^3 + 3\mu^2 + \mu,$$

$$\mu_4' = E(X^4) = \sum_{x=0}^{\infty} x^4.p(x;,\mu)$$

$$\mu_4' = (\mu^4 + 6\mu^3 + 7\mu^2 + \mu) - 4\mu(\mu^3 + 3\mu^2 + \mu) + 6\mu^2(\mu^2 + \mu) - 3\mu^4$$

$$\mu_4' = 3\mu^2 + \mu.$$

Now $$\beta_1 = \frac{\mu_{3^2}}{\mu_2^3} = \frac{\mu^2}{\mu^3} = \frac{1}{\mu},\ and$$

$$\beta_2 = \frac{\mu_4}{\mu_2^2} = \frac{3\mu^2 + \mu}{\mu^2} = 3 + \frac{1}{\mu}.$$

- The shape of the Poisson distribution depends on the value of its parameter μ. As the distribution takes on an infinite number of x values (theoretically), the distribution will be positively skewed. The distribution tends to be symmetrical as μ becomes larger and larger.

Reproductive Property

If two independent r.v. $X \, and \, Y$ have Poisson distribution with parameters μ and v, then their sum $X+Y$ has also a Poisson distribution with parameter $\mu + v$.

Poisson Process

In an earlier section, the Poisson distribution was obtained as a limiting approximation to the binomial distribution. The terms of the Poisson distribution can also be derived from a Poisson process which may be defined as a physical process governed at least in part by some random mechanism. The occurrence of traffic deaths per month in a city is an example of a Poisson process. A Poisson process has the following properties.

- The probability that an event occurs in a very short time interval Δ is proportional to the length of the time interval, i.e. is approximately λ Δ, when λ is a positive quantity and may be interpreted as the average number of occurrences per unit of time.
- The probability that two or more events occur in such a short interval is so small that it can be neglected.
- Events occurring that two or more events occur in such a short interval is so small that it can be neglected.
- Events occurring in non- overlapping intervals of time are statistically independent.

These properties are assumed to hold for events occurring randomly in regions of space, etc. Under these properties, it can be shown that the probability for the number of occurrences of a random event in an interval of stated length t is given by the Poisson distribution with the parameter λt. That is, the Poisson process formula is

$$p(x;\lambda t)=\frac{e^{-\lambda t}(\lambda t)^x}{x!},$$

Where t = number of units of time,

$x=$ number of occurrences in t units of time, and

$\lambda =$ average number of occurrences per unit of time.

Moment Generating and Cumulant Generating functions of the Poisson Distribution

The $m.g.f.$ for the Poisson distribution with respect to the origin, is found as below:

$$M_0(t)=E(e^{tx})=\sum_{x=0}^{\infty}e^{tx}\,p(x;\mu)==e^{\mu(e^t-1)}$$

The cumulant generating function *(c.g.f.)* of the Poisson distribution with respect to origin is given by

$$k(t)=\log_e M_0(t)$$

$$=\mu(e^t-1)$$

$$=\mu\left(t+\frac{t^2}{2!}+\frac{t^3}{3!}+\ldots+\frac{t^r}{r!}+\ldots\right)$$

$$k_r=co-efficient\,of\,\frac{t^r}{r!}in\,k(t)=\mu,for\,all\,r$$

Hence all the cumulant of the Poisson distribution are equal to μ.

Negative Binomial Distribution

In the binomial experiments, the number of successes varies and the number of trials is fixed. But there are experiments in which the number of successes is fixed and the number of trials varies to produce the fixed number of successes. Such experiments are called negative binomial experiments. In other words, a negative binomial experiment possesses the following properties:

- The outcomes of each trial may be classified into one of two categories: success (S) and failure (F).
- The probability of success, denoted by p, remains constant for all trials.
- The successive trials are all independent.
- The experiment is repeated a variable number of times to obtain a fixed number of successes.

When X denotes the number of trials to produce k successes in a negative binomial experiment, it is called a negative binomial variable, and its p.d., is called the negative binomial distribution. When the negative binomial r.v X assumes a value x on which the kth success occurs, the negative binomial distribution is given by

$$P(X = x) = \binom{x-1}{k-1} p^k q^{x-k}, \quad x = k, k+1, k+2, \ldots$$

The negative binomial p.d. has two parameter k and $p > 0$, and is generally denoted by $b(x; k, p)$.

Properties of the Negative Binomial Distribution

The important properties of the negative binomial distribution are given below:

- The mean of the negative binomial distribution is less than its variance.
- The negative binomial distribution is always positively skewed.

Geometric Distribution

When an experiment consists of independent trials with probability p of success and the trials are repeated until the first success occurs. It is called a geometric experiment. In other words, ageometric experiment has the following four properties:

- The outcomes of each trial may be classified into one of two categories, success and failure.
- The probability of success p remains constant for all trials.
- The successive trials are all independent.
- The experiment is repeated a variable number of times until the first success is obtained.

If X represents the number of trials needed for the first success, then X is called a geometric r.v. and its p.d. is called the geometric distribution.

It has only one parameter p and is denoted by $g(x\,;p).$ The geometrical distribution derives its name from the fact that its successive terms constitute a geometric progression. Since a geometric r.v. represents how long one has to wait for a success, it is also called a waiting time r.v. It is interesting to note that a geometric distribution is a special case of a negative binomial distribution when $k=1$.

Properties of the Geometric Distribution:

- The mean and variance of the geometric distribution are $\mu=\frac{1}{p}$ and $\sigma^2=\frac{q}{p^2}$
- The distribution is positively skewed.

Moment Generating Function of the Geometric Distribution.

The m.g.f. of the geometric distribution is derived as below:

By definition, $M_0(t)=E\left(e^{tx}\right)=\sum_{x=1}^{\infty}e^{tx}q^{x-1}p$

$$M_0(t)=\frac{pe^t}{1-qe^t}, where\,qe^t<1$$

Multinomial Distribution

A binomial experiment becomes a multinomial experiment when there are more than two possible outcomes of each trial. *For example*, manufactured items may be classified as good, average, or inferior; or a road accident may result in no injury, minor injuries, severe injuries, or fatal injuries.

A multinomial experiment has the following properties:

- The outcomes of each trial may be classified into one of k mutually exclusive categories $C_1, C_2, \ldots, C_k$.
- The probability of the ith outcome is p_i which remains constant and $\Sigma p_i = 1$.
- The successive trials are all independent.
- The experiment is repeated a fixed number of times, say n.

CHAPTER 7

CONTINUOUS PROBABILITY DISTRIBUTIONS

INTRODUCTION

In this chapter, we shall consider some important continuous probability distributions or density functions which are met in practice. Of all the continuous probability distributions, the normal distribution is perhaps the most important distribution which is used extensively in solving problems both in probability and in statistical inference.

Continuous Uniform Distribution

One of the simplest continuous distributions in all of statistics is the continuous uniform distribution. This distribution has a smooth density function and thus the probability is uniform in a closed interval, say [*A, B*]. Although its applications are not as affluent as those for other distributions, it is appropriate for the beginner to begin this introduction to continuous distributions with the uniform distribution.

The density function of a continuous r.v. X is called a uniform distribution when between the end points any two subintervals of the same length containing X, have the same probability.

Alternatively, a r.v. X is said to be uniformly distributed if its density function is defined as

$$f(x)=\frac{1}{b-a}, a\le x\le b$$

$= 0,$ *elsewhere.*

The distribution derives its name from the fact that its density is constant or uniform over the interval $[a,b]$ and is 0 elsewhere.

It is also called the rectangular distribution because its total probability is confined to a rectangular region with base equal to $(b-a)$ and height $\frac{1}{(b-a)}$. The parameter of this distribution are a and b. Since it is asserted that X with this distribution is a random variable, therefore we must have that

$$\int_{-\infty}^{\infty} f(x)dx = \int_{a}^{b} \frac{1}{b-a} dx = \frac{1}{b-a}[x]_a^b = \frac{b-a}{b-a} = 1$$

This distribution arises in the study of rounding off errors, etc. Its distribution function will be

$$F(x) = \begin{cases} 0, & for\, x < a \\ \dfrac{x-a}{b-a} & for\, a \le x \le b \\ 1 & for\, x > b, \end{cases}$$

Properties of the Uniform Distribution

- Let X have the uniform over $[a,b]$. Then its mean is $\frac{(a+b)}{2}$ and variance is $\frac{(b-a)^2}{12}$.

$$E(X) = \int_{-\infty}^{\infty} x\, f(x)dx = \frac{1}{b-a}\int_{a}^{b} x\, dx = \frac{1}{b-a}\left[\frac{x^2}{2}\right]_a^b$$

$$= \frac{b^2 - a^2}{2(b-a)}$$

$$= \frac{a+b}{2}$$

And $Var(X) = E(X^2) - [E(X)]^2$, where

$$E(X^2) = \int_a^b x^2 \left(\frac{1}{b-a} \right) dx$$

$$= \frac{1}{b-a} \int_a^b x^2 \, dx$$

$$= \frac{1}{b-a} \left[\frac{x^3}{3} \right]_a^b$$

$$= \frac{b^3 - a^3}{3(b-a)}$$

$$= \frac{a^2 + ab + b^2}{3}$$

$$\therefore \; Var(X) = E(X^2) - [E(X)]^2$$

$$= \frac{a^2 + ab + b^2}{3} - \frac{(a+b)^2}{4}$$

$$= \frac{(b-a)^2}{12}$$

- The shape of the distribution is rectangular.

Moment Generating Function of the Uniform Distribution

The m.g.f is obtained as

$$M_0(t) = E\left[e^{tX}\right] = \int_a^b e^{tx} \frac{1}{b-a} dx$$

$$= \frac{1}{b-a} \int_a^b e^{tx} \, dx$$

$$=\frac{1}{b-a}\left[\frac{e^{tx}}{t}\right]_a^b$$

$$=\frac{e^{bt}-e^{at}}{(b-a)t}$$

Putting $a=0\, and\, b=1,$ we obtain a r.v. with a uniform distribution in the interval $[0,1]$. Then its kth moment about the origin is

$$\mu_k^{'}=\int_0^1 x^k dx$$

$$=\left[\frac{x^{k+1}}{k+1}\right]_0^1=\frac{1}{k+1}$$

Example: Suppose that a large conference room at a certain company can be reserved for no

more than 4 hours. Both long and short conferences occur quite often. In fact, it can be assumed that the length *X* of a conference has a uniform distribution on the interval [0, 4].

a. What is the probability density function?

b. What is the probability that any given conference lasts at least 3 hours?

Solution: a. The appropriate density function for the uniformly distributed random variable X in this situation is

$$f(x)=\begin{cases}\frac{1}{4}, & 0\leq x\leq 4,\\ 0, & elsewhere.\end{cases}$$

b. $$P[X\geq 3]=\int_3^4\frac{1}{4}dx=\frac{1}{4}.$$

Although the normal distribution can be used to solve many problems in engineering and science, there are still numerous situations that require different types of density functions. Two such density functions, the gamma and exponential distributions, are discussed in this section. It turns out that the exponential distribution is a special case of the gamma distribution. Both find a large number of applications. The exponential and gamma distributions play an important role in both queuing theory and reliability problems. Time between arrivals at service facilities and time to failure of component parts and electrical systems often are nicely modeled by the exponential distribution. The relationship between the gamma and the exponential allows the gamma to be used in similar types of problems. More details and illustrations will be supplied later in the section.

Exponential Distribution

A random variable X is said to have an exponential distribution with parameter λ, if its p.d.f. is defined by

$$f(x) = \lambda e^{-\lambda x} \quad for\, x > 0,$$

$$= 0, elsewhere$$,

Where $\lambda > 0$. The p.d.f. may also be written as

$$f(x) = \frac{1}{\beta} e^{-\frac{x}{\beta}} \quad for\, x > 0,$$

$$= 0, \quad elsewhere.$$

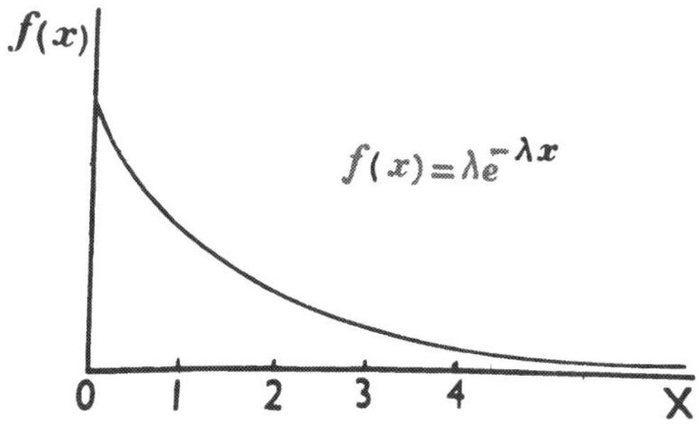

The function $f(x)$ is a proper p.d.f. since

$$\int_{-\infty}^{\infty} f(x)\,dx = \int_{0}^{\infty} \lambda e^{-\lambda x}\,dx = \left[-e^{-\lambda x}\right]_{0}^{1} = 1$$

The distribution function of the exponential r.v. X is given by
$F(x) = P(X \le x)$

$$= \int_{0}^{x} \lambda e^{-\lambda x}\,dt$$

$$= \left[-e^{-\lambda t}\right]_{0}^{x}$$

$$= 1 - e^{-\lambda x}, \textit{for}\ x > 0,$$

$$= 0, \quad \textit{elsewhere},$$

And hence $P(X > x) = e^{-\lambda x}$

Properties

The following are some of the important properties of the exponential distribution with parameter λ.

- The mean and standard deviation of the exponential distribution are equal.

Now $E(X) = \mu = \int_{-\infty}^{\infty} x\,f(x)\,dx = \int_{0}^{\infty} x.\lambda e^{-\lambda x}\,dx$

To integrate it by parts, we use the formula $\int u\,dv = uv - \int v\,du$

And make the substitution, $\lambda e^{-\lambda x} dx = dv$ and $x = u,$ so that $v = -e^{-\lambda x}$ *and* $du = dx.$ Then, we have

$$\int_{0}^{\infty} x.\lambda e^{-\lambda x}\,dx = \left[-x e^{-\lambda x}\right]_{0}^{\infty} + \int_{0}^{\infty} e^{-\lambda x}\,dx$$

$$= 0 + \left[\frac{-e^{-\lambda x}}{\lambda}\right]_{0}^{\infty} = \frac{1}{\lambda}$$

Thus $E(X)$ or $\mu = \frac{1}{\lambda}$

Again $Var(X) = E(X^2) - [E(X)]^2$,

where

$$E(X^2) = \int_0^\infty x^2 . \lambda . e^{-\lambda x}\, dx$$

$$= \lambda \left[x^2 \left(-\frac{1}{\lambda} e^{-\lambda x} \right) \right]_0^\infty + \int_0^\infty 2x . e^{-\lambda x}\, dx$$

$$= 0 + \frac{2}{\lambda} E(X) = \frac{2}{\lambda^2}$$

$$Var(X) = \sigma^2 = \frac{2}{\lambda^2} - \left(\frac{1}{\lambda} \right)^2 = \frac{1}{\lambda^2}$$

and hence $\sigma = \frac{1}{\lambda}$.

- The distribution is extremely skewed and thus there does not exist any mode.

Moment Generating Function of Exponential Distribution

The m.g.f. is obtained as follows:

$$M_0(t) = E\left[e^{tX} \right] = \int_0^\infty e^{tx} \lambda\, e^{-\lambda x} dx$$

$$= \lambda \int_0^\infty e^{-(\lambda - t)x} dx$$

$$= \lambda \left[\frac{-e^{-(\lambda - t)x}}{\lambda - t} \right]_0^\infty$$

$$=\frac{\lambda}{\lambda-t}\, for\, t<\lambda.$$

Example: The duration of long- distance telephone calls is found to be exponentially distributed with a mean of 3 minutes. What is the probability that a call will last (i) more than 3 minutes, (ii) more than 5 minutes?

Solution: Let X be the exponential r.v. with parameter λ. Then the mean, i.e. $\frac{1}{\lambda}=3,$ so that $\lambda=\frac{1}{3}.$

Now, i. the probability that a call will last more than 3 minutes, is given by

$$P(X>3)=\int_{3}^{\infty}\left(\frac{1}{3}\right)e^{-\frac{x}{3}}\,dx$$

$$=\left[-e^{-\frac{x}{3}}\right]_{3}^{\infty}=e^{-1}=0.3679$$

And (ii) the probability that a call will last more than 5 minutes, is given by

$$P(X>5)=\int_{5}^{\infty}\left(\frac{1}{3}\right)e^{-\frac{x}{3}}\,dx=\left[-e^{-\frac{x}{3}}\right]_{3}^{\infty}=e^{-1.7}=0.1827$$

The Memoryless Property and Its Effect on the Exponential Distribution

The types of applications of the exponential distribution in reliability and component or machine lifetime problems are influenced by the **memoryless** (or lack-of memory) **property** of the exponential distribution. For example, an electronic component where lifetime has an exponential distribution, the probability that the component lasts, say, t hours, that is, $P(X \geq t)$, is the same as the conditional probability

$P(X \geq t_0 + t | X \geq t_0)$.

So if the component "makes it" to t hours, the probability of lasting an additional t hours is the same as the probability of lasting t hours. There is no "punishment" through wear that may have ensued for lasting the first t hours. Thus, the exponential distribution is more appropriate when the memory less property is justified. But if the failure of the component is a result of gradual or slow wear (as in mechanical wear), then the exponential does not apply and either the gamma or the Weibull distribution may be more appropriate.

The gamma and beta distribution derives there name from the well-known gamma and beta function, which are very important in many areas of probability theory and mathematics. Therefore, before proceeding to these distributions, it is appropriate to review the gamma and beta functions and some of their main properties.

Gamma Function

The gamma function for any number $n > 0,$ denoted by $\Gamma(n),$ is defined by

$$\Gamma(n) = \int_{0}^{\infty} x^{n-1} e^{-x} dx$$

We can give an alternative meaning to $\Gamma(n),$ by proving that $\Gamma(n+1) = n!.$

Now by definition $\Gamma(n+1) = \int_{0}^{\infty} x^{n}.e^{-x}\, dx, n > 0$

To integrate the above integral by parts, we use the formula $\int_{-\infty}^{\infty} u\, dv = uv - \int_{-\infty}^{\infty} v\, du$ and make the substitutions $u = x^{n}, dv = e^{-x}dx$ so that $du = n x^{n-1}\ and\ v = -e^{-x}$. Then we obtain

$$\Gamma(n+1) = \left[-e^{-x}x^{n}\right]_{0}^{\infty} + \int_{0}^{\infty} n x^{n-1}.e^{-x}\, dx$$

$$= n \int_0^\infty x^{n-1} e^{-x} \, dx$$

$= n\Gamma(n)$ by the definition of Γ function.

Again $\Gamma(n) = \int_0^\infty x^{n-1} e^{-x} \, dx$

$$= \left[-e^{-x} x^{n-1} \right]_0^\infty + \int_0^\infty (n-1) e^{-x} x^{n-2} dx$$

$$= (n-1)\Gamma(n-1)$$

Let n be a positive integer. Then repeating the application of $\Gamma(n) = (n-1)\Gamma(n-1)$ we get

$$\Gamma(n+1) = n.(n-1)\Gamma(n-1)$$

$$= n(n-1)(n-2)\ldots3.2.\Gamma(1)$$

Putting $n = 0,$ we find that

$$\Gamma(1) = \int_0^\infty e^{-x} \, dx = \left[-e^{-x} \right]_0^\infty$$

$$= -e^{-\infty} + 1 = 1$$

Hence $\Gamma(n+1) = n(n-1)(n-2)\ldots3.2.1 = n\,!$

Writing $x = y^2$ in $\Gamma(n) = \int_0^\infty x^{n-1} e^{-x} \, dx,$ we have

$$\Gamma(n) = 2\int_0^\infty e^{-y^2} y^{2n-1} \, dy,$$

Which is another form of Γ *function.*

Putting $n = \frac{1}{2}$ in this form, we obtain

$$\Gamma\left(\frac{1}{2}\right)=2\int_{0}^{\infty} e^{-y^2}\,dy=\int_{-\infty}^{\infty} e^{-y^2}\,dy=\sqrt{\pi}\ .$$

Gamma Distribution

A continuous r.v. X is said to have a gamma distribution with parameter $m>0,$ if its p.d.f. is defined by

$$f(x)=\begin{cases}\dfrac{1}{\Gamma(m)}x^{m-1}.e^{-x} & for\ 0\le x\le\infty\\ 0, & elsewhere\end{cases}$$

A gamma variable with parameter m is usually denoted by $\gamma(m).$

A straightforward integration shows that $\int_{0}^{\infty} f(x)\,dx=1$ and hence it represents a p.d.f.

The distribution function $F(x)$ is

$$F(x)=\int_{0}^{x}\frac{1}{\Gamma(m)}x^{m-1}.e^{-x}\,dx,\ x\ge 0.$$

$$=0,\qquad x<0$$

Which is also called the incomplete Gamma function and has been tabulated by Karl Pearson.

or

The continuous random variable X has a gamma distribution, with parameters $\alpha\ and\ \beta,$ if its density function is given by

$$F(x;\alpha,\beta)=\begin{cases}\dfrac{1}{\beta^{\alpha}\,\Gamma(\alpha)}x^{\alpha}\,e^{-\frac{x}{\beta}} & ,\ x>0,\\ 0, & elsewhere\end{cases}$$

where $\alpha>0\,and\ \beta>0.$

Properties of Gamma Distribution

The important properties of the gamma distribution are given as follows:

- The mean and variance of the gamma distribution are equal to its parameter m.

Now, $\mu = E(X) = \int_{-\infty}^{\infty} x f(x) dx = \int_{0}^{\infty} x . \frac{1}{\Gamma(\mathrm{m})} x^{m-1} . e^{-x} dx$

$$= \int_{0}^{\infty} \frac{1}{\Gamma(m)} x^{m} . e^{-x} dx$$

$$= \frac{\Gamma(m+1)}{\Gamma(m)} = \frac{m\Gamma(m)}{\Gamma(m)} = m$$

And $Var(X) = E(X^2) - [E(X)]^2$, where

$$E(X^2) = \int_{0}^{\infty} x^2 \frac{1}{\Gamma(m)} x^{m-1} . e^{-x} dx$$

$$= \frac{\Gamma(m+2)}{\Gamma(m)} = \frac{(m+1)m\Gamma(m)}{\Gamma(m)} = m(m+1)$$

$$\sigma^2 = m(m+1) - m^2 = m$$

Hence mean and variance are each equal to m.

- The curve of $f(x)$ is asymptotic to the X-axis. If $m > 1$, the curve has a mode at $x = m-1$. If $m > 2$, it touches the $X-axis$ at the origin.

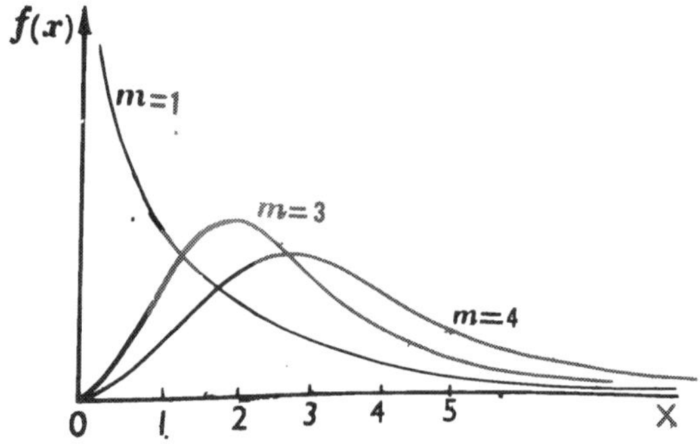

- The curve becomes asymptotic to both axes when m lies between zero and one.
- Reproductive Property. The sum of two independent Gamma distributions with parameter m and n is a Gamma distribution with parameter $(m+n)$.
- Moment Generating Function

The m.g.f. of X with respect to the origin is

$$M_0(t)=\int_0^\infty e^{tx}\frac{1}{\Gamma(m)}x^{m-1}e^{-x}\,dx$$

$$=\int_0^\infty \frac{1}{\Gamma(m)}.x^{m-1}e^{-x(1-t)}\,dx$$

Let $u=x(1-t)$, so that $dx=\frac{du}{1-t}$. Then substituting these values, we get

$$M_0(t)=\frac{1}{\Gamma(m)}\int_0^\infty e^{-u}\left(\frac{u}{1-t}\right)^{m-1}\frac{du}{1-t}$$

$$=\frac{1}{(1-t)^m}.\int_0^\infty \frac{1}{\Gamma(m)}.u^{m-1}.e^{-u}\,du$$

$=(1-t)^{-m}$, provided that $|t|<1$.

On differentiating $M_0(t)$ r times with respect to t and putting $t=0$, we find

$$\mu_r'=m(m+1)\ldots(m+r-1)$$

Thus cumulant generating function for this distribution with respect to origin is:

$$k(t)=\log_e M_0(t)=-m\log(1-t)$$

$$=m\left[t+\frac{t^2}{2}+\frac{t^3}{3}+\ldots\right]$$

$$= m\sum_{r=1}^{\infty}\frac{t^r}{r!}(r-1)!$$

Thus the rth cumulant is given as

$$k_r = m(r-1)! = m\Gamma(r)$$

In the foregoing, we provided the foundation for the application of the exponential distribution in "time to arrival" or time to Poisson event problems. We will illustrate some applications here and then proceed to discuss the role of the gamma distribution in these modeling applications. Notice that the mean of the exponential distribution is the parameter β, the reciprocal of the parameter in the Poisson distribution. The reader should recall that it is often said that the Poisson distribution has no memory, implying that occurrences in successive time periods are independent. The important parameter β is the mean time between events. In reliability theory, where equipment failure often conforms to this Poisson process, β is called mean time between failures. Many equipment breakdowns do follow the Poisson process, and thus the exponential distribution does apply. Other applications include survival times in biomedical experiments and computer response time.

In the following example, we show a simple application of the exponential distribution to a problem in reliability. The binomial distribution also plays a role in the solution.

Example: Suppose that a system contains a certain type of component whose time, in years, to failure is given by T. The random variable T is modeled nicely by the exponential distribution with mean time to failure $\beta = 5$. If 5 of these components are installed in different systems, what is the probability that at least 2 are still functioning at the end of 8 years?

Solution: The probability that a given component is still functioning after 8 years is given by

$$P(T > 8) = \frac{1}{5}\int_{8}^{\infty} e^{-\frac{t}{5}}\,dt = e^{-\frac{8}{5}} \approx 0.2.$$

$$P(X \geq 2) = \sum_{x=2}^{5} b(x;5,0.2) = 1 - \sum_{x=0}^{1} b(x;5,0.2)$$

$$= 1 - 0.7373$$

$= 0.2627$

The importance of the gamma distribution lies in the fact that it defines a family of which other distributions are special cases. But the gamma itself has important applications in waiting time and reliability theory. Whereas the exponential distribution describes the time until the occurrence of a Poisson event (or the time between Poisson events), the time (or space) occurring until a *specified number of Poisson events occur* is a random variable whose density function is described by the gamma distribution. This specific number of events is the parameter α in the gamma density function. Thus, it becomes easy to understand that when $\alpha = 1$, the special case of the exponential distribution occurs. The gamma density can be developed from its relationship to the Poisson process in much the same manner as we developed the exponential density. The details are left to the reader. The following is a numerical example of the use of the gamma distribution in a waiting-time application.

***Example*:** Suppose that telephone calls arriving at a particular switchboard follow a Poisson process with an average of 5 calls coming per minute. What is the probability that up to a minute will elapse by the time 2 calls have come in to the switchboard?

***Solution*:** The Poisson process applies, with time until 2 Poisson events following a gamma distribution with $\beta = \frac{1}{5}$ and $\alpha = 2$. Denote by X the time in minutes that transpires before 2 calls come. The required probability is given by

$$P(X \le 1) = \int_0^1 \frac{1}{\beta^2} x e^{-\frac{x}{\beta}} dx$$

$$= 25 \int_0^1 x e^{-5x} dx$$

$$= 1 - e^{-5}(1 + 5) = 0.96$$

While the origin of the gamma distribution deals in time (or space) until the occurrence of α Poisson events, there are many instances where a gamma distribution works very well even though there is no clear Poisson structure. This is particularly true for survival time problems in both engineering and biomedical applications.

Example: In a biomedical study with rats, a dose-response investigation is used to determine the effect of the dose of a toxicant on their survival time. The toxicant is one that is frequently discharged into the atmosphere from jet fuel. For a certain dose of the toxicant, the study determines that the survival time, in weeks, has a gamma distribution with $\alpha = 5$ and $\beta = 10$. What is the probability that a rat survives no longer than 60 weeks?

Solution: Let the random variable X be the survival time (time to death). The required probability is

$$P(X \le 60) = \frac{1}{\beta^5}\int_0^{60} \frac{x^{\alpha-1}e^{-y}}{\Gamma(\alpha)}\,dy$$

If we let $y = \frac{x}{\beta}$, so $x = \beta y$, we have

$$P(X \le 60) = \int_0^{60} \frac{\left(y^4 e^{-y}\right)}{\Gamma(5)}\,dy,$$

$$P(X \le 60) = 0.715.$$

Example: It is known, from previous data, that the length of time in months between customer complaints about a certain product is a gamma distribution with $\alpha = 2$ and $\beta = 4$. Changes were made to tighten quality control requirements. Following these changes, 20 months passed before the first complaint. Does it appear as if the quality control tightening was effective?

Solution: Let X be the time to the first complaint, which, under conditions prior to the changes, followed a gamma distribution with $\alpha = 2\, and\ \beta = 4$. The question centers around how rare $X \ge 20$ is, given that $\alpha\ and\ \beta$ remain at values 2 and 4, respectively. In other words, under the prior conditions is a "time to complaint" as large as 20 months reasonable?

$$P(X \ge 20) = 1 - \frac{1}{\beta^{\alpha}}\int_0^{20} \frac{x^{\alpha-1}e^{-\frac{x}{\beta}}}{\tilde{A}(\alpha)}\,dx.$$

Again using $y = \frac{x}{\beta}$, we have

$$P(X \geq 20) = 1 - \int_0^5 \frac{ye^{-y}}{\Gamma(2)} dy = 0.04,$$

As a result, we could conclude that the conditions of the gamma distribution with $\alpha = 2$ and $\beta = 4$ are not supported by the data that an observed time to complaint is as large as 20 months. Thus, it is reasonable to conclude that the quality control work was effective.

Example: Based on extensive testing, it is determined that the time Y in years before a major repair is required for a certain washing machine is characterized by the density function

$$f(y) = \begin{cases} \frac{1}{4} e^{-\frac{y}{4}}, y \geq 0, \\ 0, else\,where \end{cases}$$

Note that Y is an exponential random variable with $\mu = 4$ years. The machine is considered a bargain if it is unlikely to require a major repair before the sixth year. What is the probability $P(Y > 6)$? What is the probability that a major repair is required in the first year?

Solution: Consider the cumulative distribution function $F(y)$ for the exponential distribution,

$$F(y) = \frac{1}{\beta} \int_0^y e^{-\frac{t}{\beta}} dt = 1 - e^{-\frac{y}{\beta}}$$

Then

$$P(Y > 6) = 1 - F(6) = e^{-\frac{3}{2}} = 0.2231.$$

Thus, the probability that the washing machine will require major repair after year six is 0.223. Of course, it will require repair before year six with probability 0.777. Thus, one might conclude the machine is not really a bargain. The probability that a major repair is necessary in the first year is

$$P(Y<1)=1-e^{-\frac{1}{4}}=1-0.779=0.221.$$

Beta Function

The beta function for any two positive numbers m,n denoted by $B(m,n)$, is defined by

$$B(m,n)=\int_0^1 x^{m-1}(1-x)^{n-1}\,dx, \quad for\, m>0, n>0$$

when $m=n=1, B(1,1)=\int_0^1 x^0(1-x)^0\,dx=1$

Writing $z=1-x,$ we find that $B(m,n)=-\int_1^0 (1-z)^{m-1} z^{n-1}\,dz$

$$=\int_0^1 (1-z)^{m-1} z^{n-1}\,dz$$

$$=B(n,m)$$

Hence the $B-function$ is symmetrical about $m\, and\, n.$

Now, let $x=\sin^2\theta$ so that $dx=2\sin\theta\cos\theta\,d\theta.$ Then the substitution gives

$$B(m,n)=2\int_0^{\frac{\pi}{2}} \sin^{2m-1}\theta\cos^{2n-1}\theta\,d\theta$$

Putting $m=n=\frac{1}{2},$ we find that $B\left(\frac{1}{2},\frac{1}{2}\right)=2\int_0^{\frac{\pi}{2}} d\theta=\pi$

Again let $x=\frac{1}{1+z}\, so\, that\, dx=-\frac{1}{(1+z)^2}dz$. Then

$B(m,n)=\int_0^\infty \frac{z^{n-1}}{(1+z)^{m+n}}dz$, which is also B- function.

It is interesting that the gamma and beta functions are related according to the following formula:

$$B(m,n)=\frac{\Gamma(m).\Gamma(n)}{\Gamma(m+n)}, for\ m>0, n>0.$$

Putting $m=n=\frac{1}{2}$, we have

$$B\left(\frac{1}{2},\frac{1}{2}\right)=\frac{\Gamma\left(\frac{1}{2}\right).\Gamma\left(\frac{1}{2}\right)}{\Gamma(1)}=\left[\Gamma\left(\frac{1}{2}\right)\right]^2$$

But $B\left(\frac{1}{2},\frac{1}{2}\right)=\pi$

Hence $\Gamma\left(\frac{1}{2}\right)=\sqrt{\pi}$ or $\int_0^\infty e^{-x}.x^{-\frac{1}{2}}dx=\sqrt{\pi}$

This is a very important result.

Beta Distribution of the First Kind

A continuous r.v. X. is said to have a beta distribution with two parameter $m\ and\ n$, if its p. d. f is defined by

$$f(x)=\begin{cases}\frac{1}{B(m,n)}x^{m-1}(1-x)^{n-1}, & 0\le x\le 1; m,n>0\\ 0, & else where\end{cases}$$

This distribution is known as a beta distribution of the first kind and a beta variable of the first kind is briefly referred to as $\beta_1(m,n)$. It is a proper probability density function since the area under the curve $f(x)$ is one.

The distribution function $F(x)$ is given by

$$f(x)=\begin{cases}0, & for\, x<0\\ \int_0^x \frac{1}{B(m,n)}.x^{m-1}(1-x)^{n-1}\,dx, & for\, 0\le x\le 1\\ 0, & for\, x>1\end{cases}$$

This is called the incomplete beta function, and it has been extensively tabulated.

Properties of $\beta_1(m,n)$

The main properties of this distribution are given below:

- The mean and variance of this distribution are $\frac{m}{m+n}$ and $\frac{mn}{(m+n)^2(m+n+1)}$, respectively. They are computed as below

$$\mu=E(X)=\int_0^1 \frac{1}{B(m,n)}.x^{m-1}(1-x)^{n-1}.x\,dx$$

$$=\int_0^1 \frac{1}{B(m,n)}.x^{m}(1-x)^{n-1}\,dx$$

$$=\frac{B(m+1,n)}{B(m,n)}$$

$$=\frac{m\Gamma(m)\Gamma(n)}{(m+n)\Gamma(m+n)}.\frac{\Gamma(m+n)}{\Gamma(m).\Gamma(n)}=\frac{m}{m+n}$$

$Var(X)=\sigma^2=E(X^2)-[E(X)]^2$, where

$$E(X^2)=\int x^2 f(x)\,dx$$

$$=\int_0^1 \frac{1}{B(m,n)}.x^{m+1}(1-x)^{n-1}\,dx$$

$$=\frac{B(m+2,n)}{B(m,n)}=\frac{m(m+1)}{(m+n)(m+n+1)}$$

$$\sigma^2=\frac{m(m+1)}{(m+n)(m+n+1)}-\left(\frac{m}{m+n}\right)^2$$

$$=\frac{mn}{(m+n)^2(m+n+1)}$$

Higher Moments. The rth moment about the origin 0 is given by

$$\mu_r'=E(X^r)=\frac{1}{B(m,n)}\int_0^1 x^r.x^{m-1}(1-x)^{n-1}\,dx$$

$$=\frac{B(m+r,n)}{B(m,n)}=\frac{\Gamma(m+r)}{\Gamma(m)}\cdot\frac{\Gamma(m+n)}{\Gamma(m+r+n)}$$

$$=\frac{m(m+1)\ldots(m+r-1)}{(m+n)(m+n+1)\ldots(m+n+r-1)}$$

It should be noted that the m.g.f. for this distribution does not have a simple form

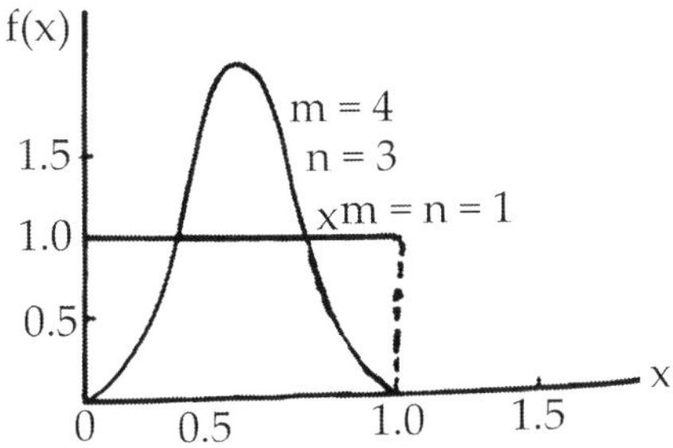

- The shape of the beta distribution for $m=4, n=3$ is indicated in the figure.
- If m and n are both greater than 1, it has a modal value at $x=\frac{m-1}{m+n-2}$. If $m=n=1$, it reduces to the uniform

distribution over the unit interval. The curve of $f(x)$ touches the X-axis at $x = 0, when\, m > 2.$

Beta distribution of Second Kind

A continuous r.v. is said to have a beta distribution of the second kind with parameter $m\, and\, n,$ and its p.d.f. is defined by

$$f(x) = \begin{cases} \frac{1}{B(m,n)} \cdot \frac{x^{m-1}}{(1+x)^{m+n}}, & for\, 0 \le x \le \infty, m,n > 0, \\ 0, & otherwise \end{cases}$$

A beta variable of the second kind is generally denoted by $\beta_2(m,n)$.

To check that the function $f(x)$ represents a proper probability distribution, we observe that

$$\int_0^\infty \frac{1}{B(m,n)} \cdot \frac{x^{m-1}}{(1+x)^{m+n}} dx$$

Let $1 + x = \frac{1}{y}$ so that $x = \frac{1-y}{y}$ and $dx = -\frac{dy}{y^2}.$ Then substitution gives

$$\int_0^\infty f(x)dx = \int_0^1 \frac{1}{B(m,n)} .y^{n-1}(1-y)^{m-1} dy = 1$$

Properties of $\beta_2(m,n)$.

The important properties of beta distribution of the second kind are given below:

- Moments about the origin are easily calculated as follows

$$\mu_r' = E(X^r) = \int_0^\infty \frac{1}{B(m,n)} \cdot \frac{x^{m-1}}{(1+x)^{m+n}} .x^r\, dx$$

$$= \frac{1}{B(m,n)} \int_0^1 y^{n-r-1}(1-y)^{m+r-1}\, dy, on\ putting\ 1+x=\frac{1}{y}$$

$$= \frac{1}{B(m,n)} \int_0^1 z^{m+r-1}(1-z)^{n+r-1}\, dz,\ where\ z=1-y$$

$$= \frac{B(m+r,n-r)}{B(m,n)} = \frac{m(m+1)\ldots(m+r-1)}{(n-1)(n-2)\ldots(n-r)}, r<n,$$

If $m>1$, the distribution is unimodal with a mode at $x=\frac{m-1}{n+1}$.

If $m=1$, the distribution is $J-shaped$.

- The curve of $f(x)$ is asymptotic to the $X-axis$ and it touches it at the origin if $m>2$.
- The curve touches the $y-axis$ at the origin if $1<m<2$, and the curve becomes asymptotic to both axes when m lies between 0 and 1.

Chi-Squared Distribution

Another very important special case of the gamma distribution is obtained by letting $\alpha=\frac{v}{2}$ and $\beta=2$, where *v* is a positive integer. The result is called the chi-squared distribution. The distribution has a single parameter, *v*, called the degrees of freedom.

The continuous random variable *X* has a chi-squared distribution, with *v* degrees of freedom, if its density function is given by

$$f(x;v)=\begin{cases} \frac{1}{2^{\frac{v}{2}}\Gamma\left(\frac{v}{2}\right)} x^{\frac{v}{2}-1} e^{-\frac{x}{2}}, & x>0, \\ 0, & elsewhere. \end{cases}$$

Where v is a positive integer.

The chi-squared distribution plays a vital role in statistical inference.

It has considerable applications in both methodology and theory. The chi-squared distribution is an important component of statistical hypothesis testing and estimation. Topics dealing with sampling distributions, analysis of variance, and nonparametric statistics involve extensive use of the chi-squared distribution.

Mean and Variance

The mean and variance of the chi-squared distribution are

$\mu = v$

$\sigma^2 = 2v$

Lognormal Distribution

The lognormal distribution is used for a wide variety of applications. The distribution applies in cases where a natural log transformation results in a normal distribution.

The continuous random variable X has a lognormal distribution if the random variable $Y = \ln(X)$ has a normal distribution with mean μ and standard deviation σ.

The resulting density function of X is

$$f(x;\mu,\sigma) = \begin{cases} \frac{1}{\sqrt{2\pi}\sigma x} e^{-\frac{1}{2\sigma^2}[\ln(x)-\mu]^2}, & x \geq 0 \\ 0, & x < 0 \end{cases}$$

Mean and Variance

The mean and variance of the lognormal distribution are

$\mu = e^{\mu+\frac{\sigma^2}{2}}$ *and* $\sigma^2 = e^{2\mu+\sigma^2}\left(e^{\sigma^2}-1\right)$.

The cumulative distribution function is quite simple due to its relationship to the normal distribution. The use of the distribution function is illustrated by the following example.

***Example*:** Concentrations of pollutants produced by chemical plants historically are known to exhibit behavior that resembles a lognormal distribution. This is important when one considers issues regarding compliance with government regulations. Suppose it is assumed that the concentration of a certain pollutant, in parts per million, has a lognormal distribution with parameters $\mu = 3.2$ and $\sigma = 1$. What is the probability that the concentration exceeds 8 parts per million?

***Solution*:** Let the random variable X be pollutant concentration. Then

$$P(X > 8) = 1 - P(X \leq 8)$$

$$P(X \leq 8) = \Phi\left[\frac{\ln(8) - 3.2}{1}\right] = \Phi(-1.12) = 0.1314.$$

Here, we use Φ to denote the cumulative distribution function of the standard normal distribution. As a result, the probability that the pollutant concentration exceeds 8 parts per million is 0.1314.

***Example*:** The life, in thousands of miles, of a certain type of electronic control for locomotives has an approximately lognormal distribution with $\mu = 5.149$ and $\sigma = 0.737$. Find the 5th percentile of the life of such an electronic control.

***Solution*:** From Table, we know that $P(Z < -1.645) = 0.05$. Denote by X the life of such an electronic control. Since $\ln(X)$ has a normal distribution with mean $\mu = 5.149$ and $\sigma = 0.737$, the 5th percentile of X can be calculated as $ln(x) = 5.149 + (0.737)(-1.645) = 3.937$.

Hence, $x = 51.265$. This means that only 5% of the controls will have lifetimes less than $51,265$ miles.

Weibull Distribution

Modern technology has enabled engineers to design many complicated systems whose operation and safety depend on the reliability of the various components making up the systems. For example, a fuse may burn out, a steel column may buckle, or a heat-sensing device may fail. Identical components subjected to identical environmental conditions will fail at different and unpredictable times. We have seen the role that the gamma and exponential distributions play in these types of problems.

Another distribution that has been used extensively in recent years to deal with such problems is the **Weibull distribution**, introduced by the Swedish physicist Waloddi Weibull in 1939.

The continuous random variable X has a Weibull distribution, with parameters α and β, if its density function is given by

$$f(x;\alpha,\beta)=\begin{cases}\alpha\,\beta\, x^{\beta-1}e^{-\alpha}x^{\beta}, & x>0\\ 0, & else\,where,\end{cases}$$

where $\alpha > 0\, and\; \beta > 0$.

Mean and variance:

The mean and variance of the Weibull distribution are

$$\mu=\alpha^{-\frac{1}{\beta}}\Gamma\left(1+\frac{1}{\beta}\right)$$

$$\sigma^2=\alpha^{-\frac{2}{\beta}}\left\{\Gamma\left(1+\frac{2}{\beta}\right)-\left[\Gamma\left(1+\frac{1}{\beta}\right)\right]^2\right\}$$

Like the gamma and exponential distributions, the Weibull distribution is also applied to reliability and life-testing problems such as the time to failure or life length of a component, measured from some specified time until it fails. Let us represent this time to failure by the continuous random variable T, with probability density function $f(t)$, where $f(t)$ is the Weibull distribution. The Weibull distribution has inherent flexibility in that it does not require the lack of memory property of the exponential distribution. The cumulative distribution function (cdf) for the Weibull can be written in closed form and certainly is useful in computing probabilities.

The cumulative distribution function for the Weibull distribution is given by

$$F(x)=1-e^{-\alpha\, x^{\beta}},\quad for\; x\geq 0,$$

for $\alpha > 0\, and\; \beta > 0$.

Summary and Discussion

In this chapter we discussed about important continuous distribution, which play very important role in probability theory.

Continuous Uniform Distribution

One of the simplest continuous distributions in all of statistics is the continuous uniform distribution. The density function of a continuous r.v. X is called a uniform distribution when between the end points any two subintervals of the same length containing X, have the same probability. Alternatively, a r.v. X is said to be uniformly distributed if its density function is defined as

$$f(x) = \frac{1}{b-a}, a \le x \le b$$

$$= 0, \textit{ elsewhere}$$

The distribution derives its name from the fact that its density is constant or uniform over the interval $[a,b]$ and is 0 elsewhere.

It is also called the rectangular distribution because its total probability is confined to a rectangular region with base equal to $(b-a)$ and height $\frac{1}{(b-a)}$. The parameters of this distribution are a and b.

This distribution arises in the study of rounding off errors, etc. Its distribution function will be

$$F(x) = \begin{cases} 0, & for\ x < a \\ \dfrac{x-a}{b-a} & for\ a \le x \le b \\ 1 & for\ x > b, \end{cases}$$

Properties of the Uniform Distribution

- Let X have the uniform over $[a,b]$. Then its mean is $\frac{(a+b)}{2}$

and variance is $\frac{(b-a)^2}{12}$.

- The shape of the distribution is rectangular.

Exponential Distribution

A random variable X is said to have an exponential distribution with parameter λ, if its p. d.f. is defined by

$$f(x) = \lambda e^{-\lambda x} \quad for\, x > 0,$$

$$= 0, elsewhere$$

where $\lambda > 0$. The p.d.f. may also be written as

$$f(x) = \frac{1}{\beta} e^{-\frac{x}{\beta}} \quad for\, x > 0,$$

$$= 0, \qquad elsewhere$$

Properties.

The following are some of the important properties of the exponential distribution with parameter λ.

- The mean and standard deviation of the exponential distribution are equal.
- The distribution is extremely skewed and thus there does not exist any mode.

Gamma Function

The gamma function for any number $n > 0$, denoted by $\tilde{A}(n)$, is defined by

$$\Gamma(n) = \int_0^\infty x^{n-1} e^{-x} dx$$

We can give an alternative meaning to $\Gamma(n)$ by proving that

$\Gamma(n+1) = n!.$

Now by definition $\Gamma(n+1) = \int_0^\infty x^n . e^{-x}\, dx, n > 0$

Gamma Distribution

A continuous r.v. X is said to have a gamma distribution with parameter $m > 0,$ if its p.d.f. is defined by

$$f(x) = \begin{cases} \frac{1}{\Gamma(m)} x^{m-1} . e^{-x} & for\ 0 \le x \le \infty \\ 0, & elsewhere \end{cases}$$

A gamma variable with parameter m is usually denoted by $\gamma(m).$

A straightforward integration shows that $\int_0^\infty f(x)\,dx = 1$ and hence it represents a p.d.f.

The distribution function $F(x)$ is

$$F(x) = \int_0^x \frac{1}{\Gamma(m)} x^{m-1} . e^{-x}\, dx,\ x \ge 0.$$

$$= 0, \qquad x < 0$$

Which is also called the incomplete Gamma function and has been tabulated by Karl Pearson.

The continuous random variable X has a gamma distribution, with parameters $\alpha\ and\ \beta,$ if its density function is given by

$$F(x;\alpha,\beta) = \begin{cases} \frac{1}{\beta^\alpha \Gamma(\alpha)} x^\alpha e^{-\frac{x}{\beta}} & , x > 0, \\ 0, & else where \end{cases}$$

where $\alpha > 0\ and\ \beta > 0.$

Properties of Gamma Distribution

The important properties of the gamma distribution are given as follows:

- The mean and variance of the gamma distribution are equal to its parameter m.
- The curve of $f(x)$ is asymptotic to the X-axis. If $m>1,$ the curve has a mode at $x=m-1$. If $m>2,$ it touches the $X-axis$ at the origin.
- The curve becomes asymptotic to both axes when m lies between zero and one.
- Reproductive Property. The sum of two independent Gamma distributions with parameter m and n is a Gamma distribution with parameter $(m+n)$.

Beta Distribution of the First Kind

A continuous r.v. X is said to have a beta distribution with two parameter $m\, and\, n,$ if its p.d.f. is defined by

$$f(x)=\begin{cases} \frac{1}{B(m,n)}x^{m-1}(1-x)^{n-1}, & 0\le x\le 1; m,n>0 \\ 0, & elsewhere \end{cases}$$

This distribution is known as a beta distribution of the first kind and a beta variable of the first kind is briefly referred to as $\beta_1(m,n)$. It is a proper probability density function since the area under the curve $f(x)$ is one.

The distribution function $F(x)$ is given by

$$f(x)=\begin{cases} 0, & for\, x<0 \\ \int_0^x \frac{1}{B(m,n)}.x^{m-1}(1-x)^{n-1}\,dx, & for\, 0\le x\le 1 \\ 0, & for\, x>1 \end{cases}$$

This is called the incomplete beta function, and it has been extensively tabulated.

Properties of $\beta_1(m,n)$

The main properties of this distribution are given below:

The mean and variance of this distribution are $\frac{m}{m+n}$ and $\frac{mn}{(m+n)^2(m+n+1)}$, respectively.

If $m\, and\, n$ are both greater than 1, it has a modal values at $x = \frac{m-1}{m+n-2}$.

If $m = n = 1$, it reduces to the uniform distribution over the unit interval.

The curve of $f(x)$ touches the X-axis at $x = 0, when\, m > 2$.

Beta distribution of Second kind

A continuous r.v. X is said to have a beta distribution of the second kind with parameter $m\, and\, n$, its p.d.f is defined by

$$f(x) = \begin{cases} \frac{1}{B(m,n)} \cdot \frac{x^{m-1}}{(1+x)^{m+n}}, & for\, 0 \le x \le \infty, m, n > 0, \\ 0, & otherwise \end{cases}$$

A beta variable of the second kind is generally denoted by $\beta_2(m,n)$.

To check that the function $f(x)$ represents a proper probability distribution, we observe that

$$\int_0^\infty \frac{1}{B(m,n)} \cdot \frac{x^{m-1}}{(1+x)^{m+n}} dx$$

Let $1 + x = \frac{1}{y}$ so that $x = \frac{1-y}{y}$ and $dx = -\frac{dy}{y^2}$. Then substitution gives

$$\int_0^\infty f(x)\,dx = \int_0^1 \frac{1}{B(m,n)} .y^{n-1}(1-y)^{m-1}\,dy = 1$$

Properties of $\hat{a}_2(m,n)$.

The important properties of beta distribution of the second kind are given below:

- Moments about the origin are easily calculated as follows

$$\mu_r' == \frac{B(m+r,n-r)}{B(m,n)} = \frac{m(m+1)\ldots(m+r-1)}{(n-1)(n-2)\ldots(n-r)}, r < n,$$

- If $m > 1,$ the distribution is unimodal with a mode at $x = \frac{m-1}{n+1}.$
  If $m = 1,$ the distribution is $J - shaped.$

- The curve of $f(x)$ is asymptotic to the $X - axis$ and it touches it at the origin if $m > 2.$

- The curve touches the $y - axis$ at the origin if $1 < m < 2,$ and the curve becomes asymptotic to both axes when m lies between 0 and 1.

Chi-Squared Distribution

The continuous random variable *X* has a chi-squared distribution, with *v* degrees of freedom, if its density function is given by

$$f(x;v) = \begin{cases} \dfrac{1}{2^{\frac{v}{2}}\Gamma\left(\dfrac{v}{2}\right)} x^{\frac{v}{2}-1} e^{-\frac{x}{2}}, & x > 0, \\ 0, & elsewhere. \end{cases}$$

where v is a positive integer.

The mean and variance of the chi-squared distribution are

$$\mu = v$$

$$\sigma^2 = 2v$$

Lognormal Distribution

The continuous random variable X has a lognormal distribution if the random variable $Y = \ln(X)$ has a normal distribution with mean μ and standard deviation σ.

The resulting density function of X is

$$f(x;\mu,\sigma)=\begin{cases}\dfrac{1}{\sqrt{2\pi}\sigma x}e^{-\frac{1}{2\sigma^2}\left[\ln(x)-\mu\right]^2}, & x\geq 0\\ 0, & x<0\end{cases}$$

Mean and Variance

The mean and variance of the lognormal distribution are

$$\mu = e^{\mu+\frac{\sigma^2}{2}}\ and\ \sigma^2 = e^{2\mu+\sigma^2}\left(e^{\sigma^2}-1\right).$$

CHAPTER 8

NORMAL DISTRIBUTION

INTRODUCTION

The normal probability distribution, which is considered the corner stone of the modern statistical theory, was discovered by Abraham de Moivre (1667-1750) as the limiting form of the binomial distribution by increasing , the number of trials, to a very large number for a fixed value of But his paper remained unnoticed until 1924 when it was found in a library by Karl Pearson (1857-1936). The name of Pierre S. Laplace (1749-1827) is also associated with the derivation of the normal distribution.

The normal distribution is also called the Gaussian distribution in honor of the great German mathematician Carl F. Gauss (1777 - 1827), who also derived its equation mathematically as the probability distribution of the errors of measurements. It was Karl Pearson who in 1893 called it the normal distribution and is best known by this name today.

A normal distribution is defined by the p.d.f.

$$f(x) = \frac{1}{\sigma\sqrt{2\pi}} e^{-\left[\frac{(x-\mu)^2}{2\sigma^2}\right]}, \; for - \infty < x \langle \infty, and \; \sigma \rangle 0$$

Where μ is the mean, σ is the standard deviation and $\pi = 3.1416$ and $e = 2.7183$ are constants. Obviously a normal distribution is characterized by two parameters μ and σ, its mean and standard deviation. Since $\int_{-\infty}^{\infty} f(x)\,dx = 1$, the function $f(x)$ is a proper probability density function. The normal distribution having mean μ and variance σ^2 is usually denoted by $N(\mu,\sigma^2)$. Thus "X is $N(\mu,\sigma^2)$" means that a r.v. X is normally distributed with mean μ and variance σ^2. The graph of the normal distribution, which is a symmetrical bell-shaped curve, is called the normal curve. The location and shape of the normal curve are determined by μ *and* σ. To put it differently, μ changes the position of the normal curve along horizontal axis while σ determines the horizontal spread. A sketch for different values of μ and σ is given below:

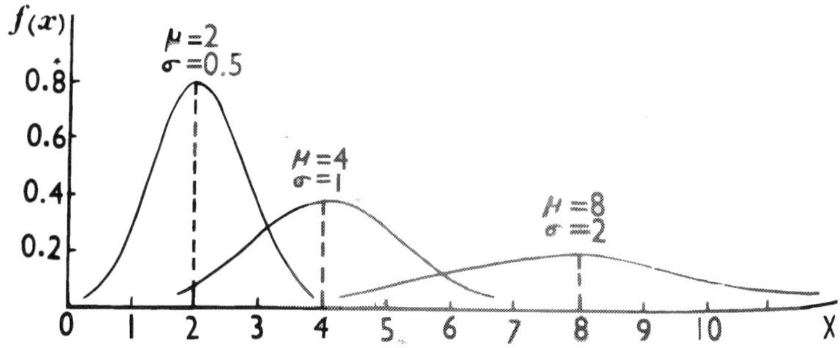

The distribution function of the normal probability distribution is given by

$$F(x) = P(X < x) = \frac{1}{\sigma\sqrt{2\pi}} \int_{-\infty}^{\infty} e^{-\left[\frac{(t-\mu)^2}{2\sigma^2}\right]} dt,$$

which is sketched below:

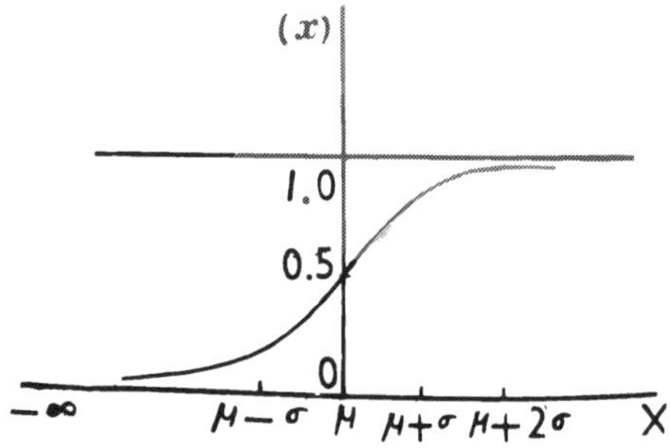

Characteristics of Normal Distribution

- One characteristic of the normal distribution is that the mean and variance are independent. There is no association between them. Knowing one tells us nothing about the value of the other. This characteristic makes the normal distribution unusual.
- The normal distribution is symmetric.
- The normal distribution is unbounded both above and below. Hence the normal distribution is defined for all real numbers.
- Its importance stems from the central limit theorem.

  –if what we observe in nature is the result of adding up lots of independent things, then the distribution of this sum tends to look normal the more things we add up.

  –As a result, sample means tend to have a normal distribution when the sample size is big enough because in calculating a mean we add things up.
- Even if the response is a discrete random variable, like a count, a normal distribution may be an adequate approximation for its distribution if we're dealing with a large sample and the values we've obtained are far removed from any boundary conditions. On the other hand count data with lots of zero values cannot possibly have a normal distribution or be transformed to approximate normality.

Standardized Normal Distribution.

A normal probability distribution depends on the values of the parameter $\mu \, and \, \sigma^2$ and the various possible values for these two parameters will result in an unlimited number of different normal distribution. The r.v. $Z = \dfrac{X - \mu}{\sigma}$, as we have seen, has zero mean and unit variance. Every normally distributed r.v. X with mean $\mu \, and \, variance = \sigma^2$ is therefore conveniently transformed into a new normal r.v Z with zero mean and unit variance by using the following expression

$$Z = \frac{X - \mu}{\sigma}$$

Then the p . d . f of Z , denoted by $\phi(z)=\frac{1}{\sqrt{2\pi}}e^{-\frac{z^2}{2}},\ for -\infty < z < \infty.$

The following figures illustrate the transformation of the original normal distributions into the standard normal distribution

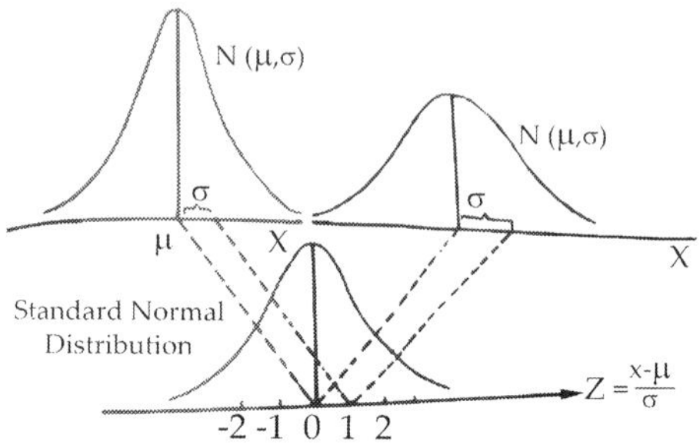

The normal probability distribution of Z which has zero mean and unit variance, is called the standardized normal distribution or unit normal distribution and is denoted by $N(0,1)$. The distribution function of the standard normal distribution, usually denoted by $\text{Ö}(z)$.

$$\text{Ö}(z)=P(Z<z)=\frac{1}{\sqrt{2\pi}}\int_{-\infty}^{z}e^{-\frac{t^2}{2}}dt,$$

which has been tabulated for positive values of z. The values of $\text{Ö}(z)$ for negative values of z are obtained from the identity $\text{Ö}(-z)=1-\text{Ö}(z)$. It should be noted that $F(x)=\text{Ö}\left(\frac{x-\mu}{\sigma}\right)$ and for any $a\ and\ b$ (positive or negative) $P(a<Z<b)=\text{Ö}(b)-\text{Ö}(a)$.

Properties of Normal Distribution

The main properties of the normal distribution are given below:

- The function $f(x)$ defining the normal distribution is a proper p.d.f. i.e. $f(x)\geq 0$ and the total area under the normal curve is unity.

Proof. Clearly $f(x)$ is always non-negative, and the total probability is

$$\int_{-\infty}^{\infty} f(x)\,dx = \int_{-\infty}^{\infty} \frac{1}{\sigma\sqrt{2\pi}} e^{\left[-\frac{(x-\mu)^2}{2\sigma^2}\right]} dx$$

Let $z = \dfrac{x-\mu}{\sigma}$. Then $\sigma\,dz = dx.$ Substituting these values, we get

$$Area = \frac{1}{\sigma\sqrt{2\pi}} e^{-\frac{z^2}{2}} \sigma\,dz$$

$$= \frac{1}{\sqrt{2\pi}}\left[\int_{-\infty}^{0} e^{-\frac{z^2}{2}} dz + \int_{0}^{\infty} e^{-\frac{z^2}{2}} dz\right]$$

The function $\int_{-\infty}^{0} e^{-\frac{z^2}{2}} dz$ being an even function of z, can be by letting $w = -z$, written as

$\int_{-\infty}^{0} e^{-\frac{z^2}{2}} dz = \int_{0}^{\infty} e^{-\frac{w^2}{2}} dw.$ Then

$$Area = 2.\frac{1}{\sqrt{2\pi}} \int_{0}^{\infty} e^{-\frac{z^2}{2}} dz$$

Let $v = \dfrac{z^2}{2},$ so that $dv = z\,dz.$ Then

$$Area = \sqrt{\frac{2}{\pi}} \int_{0}^{\infty} e^{-v} \frac{dv}{\sqrt{2v}}$$

$$= \frac{1}{\sqrt{\pi}} \int_{0}^{\infty} e^{-v} v^{-\frac{1}{2}} dv$$

$$= \frac{1}{\sqrt{\pi}} \tilde{A}\left(\frac{1}{2}\right) = \frac{1}{\sqrt{\pi}} . \sqrt{\pi} = 1$$

Thus the total area under the normal curve is unity and hence the function $f(x)$ defines a proper p.d.f.

- The mean and variance of the normal distribution are $\mu \, and \, \sigma^2$ respectively.

Proof: By definition, $\mu = E(X) = \int_{-\infty}^{\infty} x f(x) dx$

$$= \frac{1}{\sigma\sqrt{2\pi}} \int_{-\infty}^{\infty} x.e^{-\left[\frac{(x-\mu)^2}{2\sigma^2}\right]} dx$$

Let $z = \frac{x-\mu}{\sigma}$, Then $x = \mu + z\sigma \, and \, dx = \sigma dz$

when $x \to \infty, z \to \infty \, and \, x \to -\infty, z \to -\infty$

Therefore, $E(X) = \frac{1}{\sqrt{2\pi}} \int_{-\infty}^{\infty} (\mu + z\sigma) e^{-\frac{z^2}{2}} dz$

$$= \frac{\mu}{\sqrt{2\pi}} \int_{-\infty}^{\infty} e^{-\frac{z^2}{2}} dz + \frac{\sigma}{\sqrt{2\pi}} \int_{-\infty}^{\infty} z \, e^{-\frac{z^2}{2}} dz$$

The first integral represents μ times the area under a normal curve with zero mean and unit variance and hence is equal to μ. The second integral being an odd function, equals zero. Thus,

$E(X) = \mu$, i.e. μ is the mean of the normal distribution

And $Var(X) = E(X-\mu)^2 = \int_{-\infty}^{\infty} (x-\mu)^2 f(x) dx$

$$= \frac{1}{\sigma\sqrt{2\pi}} \int_{-\infty}^{\infty} (x-\mu)^2 .e^{-\left[\frac{(x-\mu)^2}{2\sigma^2}\right]} dx$$

$$= \frac{1}{\sqrt{2\pi}} \int_{-\infty}^{\infty} z^2 \sigma^2 . e^{-\frac{z^2}{2}} dz \text{ on putting } z = \frac{x-\mu}{\sigma}$$

$$= \frac{\sigma^2}{\sqrt{2\pi}} \int_{-\infty}^{\infty} z^2 . e^{-\frac{z^2}{2}} dz$$

To integrate by parts, we use the formula $\int_{-\infty}^{\infty} u\, dv = uv - \int_{-\infty}^{\infty} v\, du$ and make the substitution $dv = z e^{-\frac{z^2}{2}} dz$ and $\mu = z$ so that $v = e^{-\frac{z^2}{2}}$ and $du = dz$. Then

$$Var(X) = \frac{\sigma^2}{\sqrt{2\pi}} \left[-z . e^{-\frac{z^2}{2}} \right]_{-\infty}^{\infty} + \frac{\sigma^2}{\sqrt{2\pi}} \int_{-\infty}^{\infty} e^{-\frac{z^2}{2}} dz$$

$$= 0 + \sigma^2 = \sigma^2$$

Hence $E(X) = \mu \text{ and } Var(X) = \sigma^2$.

- The median and the mode of the normal distribution are each equal to μ, the mean of the distribution.

Now, the median, a, is given by $\int_{-\infty}^{a} f(x)\,dx = \frac{1}{2}$. Therefore

$$\frac{1}{\sigma\sqrt{2\pi}} \int_{-\infty}^{a} e^{-\left[\frac{(x-\mu)^2}{2\sigma^2}\right]} dx = \frac{1}{2}$$

or

$$\frac{1}{\sqrt{2\pi}} \int_{-\infty}^{\frac{a-\mu}{\sigma}} e^{-\frac{z^2}{2}} dz = \frac{1}{2}, \text{ on putting } z = \frac{x-\mu}{\sigma}$$

But we know from the symmetry of the standard normal distribution that

$$\frac{1}{\sqrt{2\pi}} \int_{-\infty}^{0} e^{-\frac{z^2}{2}} dz = \frac{1}{\sqrt{2\pi}} \int_{0}^{\infty} e^{-\frac{z^2}{2}} dz = \frac{1}{2}$$

$\frac{a-\mu}{\sigma}=0 \, or \, a=\mu$ i.e. μ is the median of the distribution.

Again the mode, if any, is that value of x for which $f'(x)=0$ and $f''(x)<0$.

Now $$f'(x)=\frac{1}{\sigma\sqrt{2\pi}}e^{-\left[\frac{(x-\mu)^2}{2\sigma^2}\right]}\left(\frac{x-\mu}{\sigma^2}\right)\left(-\frac{2}{2}\right)$$

$$=-\frac{1}{\sigma^3\sqrt{2\pi}}(x-\mu)e^{-\frac{(x-\mu)^2}{2\sigma^2}}$$

Equating $f'(x)=0,$ we see that $x=\mu.$

Again differentiating, we obtain

$$f'(x)=-\frac{1}{\sigma^3\sqrt{2\pi}}\left[e^{-\frac{(x-\mu)^2}{2\sigma^2}}-e^{-\frac{(x-\mu)^2}{2\sigma^2}}\frac{(x-\mu)^2}{\sigma^2}\right]$$

$$=-\frac{1}{\sigma^3\sqrt{2\pi}}e^{-\frac{(x-\mu)^2}{2\sigma^2}}\left[1-\frac{(x-\mu)^2}{\sigma^2}\right]$$

Substituting $x=\mu,$ in $f''(x),$ we see that $f'(x)=0.$ Thus $x=\mu$ is the mode of the normal distribution.Hence, the median and mode are both equal to $\mu.$ Since mean = median = mode, the normal distribution is symmetrical and unimodal.

- The mean deviation of the normal distribution is approximately $\frac{4}{5}$ of its standard deviation.

Proof. Now the mean deviation of the normal random variable X from the mean μ is given by

$$M.D = E\left[|x-\mu|\right] = \int_{-\infty}^{\infty} |x-\mu| \frac{1}{\sigma\sqrt{2\pi}} e^{-\frac{(x-\mu)^2}{2\sigma^2}} dx$$

$$= \frac{\sigma}{\sqrt{2\pi}} \int_{-\infty}^{\infty} |z| e^{-\frac{z^2}{2}} dz, \quad where\, z = \frac{x-\mu}{\sigma}$$

$$= \frac{\sigma}{\sqrt{2\pi}} \left\{ \int_{-\infty}^{0} -z e^{-\frac{z^2}{2}} dz + \int_{0}^{\infty} z e^{-\frac{z^2}{2}} dz \right\}$$

$$= \frac{2\sigma}{\sqrt{2\pi}} \int_{0}^{\infty} z e^{-\frac{z^2}{2}} dz$$

$$= \sigma\sqrt{\frac{2}{\pi}} \left[-e^{-\frac{z^2}{2}} \right]_0^{\infty}$$

$$= \sigma\sqrt{\frac{2}{\pi}}$$

$$= 0.7979\sigma$$

$$= \frac{4}{5}\sigma,$$ approximately.

- The normal curve has points of inflection which are equidistant from the mean.

Proof: The point of infection by which we mean a point at which the concavity changes, is obtained by solving the equation $f''(x) = 0.$

Differentiating the function $f(x) = \frac{1}{\sigma\sqrt{2\pi}} e^{-\frac{(x-\mu)^2}{2\sigma^2}},$ we get

$$f'(x) = \frac{-1}{\sigma^3\sqrt{2\pi}} (x-\mu) e^{-\frac{(x-\mu)^2}{2\sigma^2}}$$

Equating $f'(x)$ to zero, we see that $x=\mu.$ We also observe that $f'(x)>0\ for\ x\langle\mu\ and\ for\ x\rangle\mu, f'(x)<0.$ Thus, the maximum of the function $f(x)$ is at $x=\mu$ and its value is $\frac{1}{\sigma\sqrt{2\pi}}.$

To find the points of inflection, we take the second derivative. Thus

$$f''(x)=\frac{-1}{\sigma^3\sqrt{2\pi}}\left[-\frac{(x-\mu)^2}{\sigma^2}e^{-\frac{(x-\mu)^2}{2\sigma^2}}+e^{-\frac{(x-\mu)^2}{2\sigma^2}}\right]$$

$$=\frac{-1}{\sigma^3\sqrt{2\pi}}e^{-\frac{(x-\mu)^2}{2\sigma^2}}\left[1-\frac{(x-\mu)^2}{\sigma^2}\right]$$

Equating $f''(x)$ to zero, we are left with the following equation

$$1-\frac{(x-\mu)^2}{\sigma^2}=0$$

Which gives $(x-\mu)^2=\sigma^2$

or $x-\mu=\pm\sigma\ or\ x=\mu+\sigma\ or\ \mu-\sigma.$

At these two points, the value of the function $f(x)$ is $\frac{1}{\sigma\sqrt{2\pi e}}.$

Hence the two points of inflection of normal curve are $\left[\mu-\sigma,\frac{1}{\sigma\sqrt{2\pi e}}\right] and \left[\mu+\sigma,\frac{1}{\sigma\sqrt{2\pi e}}\right].$ In other words, the points of inflection occur on the right and on the left of the mean at a distance equal to standard deviation and thus the graph of the normal curve is bell-shaped.

- For the normal distribution, the odd order moments about the mean are all zero and the even order moments are given by

$\mu_{2n}=(2n-1)(2n-3)\ldots5.3.1\sigma^{2n}.$

Proof. The odd order moments about the mean are given by

$$\mu_{2n+1} = E(X-\mu)^{2n+1}$$

$$= \int_{-\infty}^{\infty} (x-\mu)^{2n+1} \cdot \frac{1}{\sigma\sqrt{2\pi}} e^{-\frac{(x-\mu)^2}{2\sigma^2}} dx$$

$$= \int_{-\infty}^{\infty} (\sigma z)^{2n+1} \frac{1}{\sigma\sqrt{2\pi}} e^{-\frac{z^2}{2}} \sigma dz, \text{where } z = \frac{x-\mu}{\sigma}$$

$$= \frac{\sigma^{2n+1}}{\sqrt{2\pi}} \int_{-\infty}^{\infty} z^{2n+1} . e^{-\frac{z^2}{2}} dz,$$

$= 0,$ because the integral is an odd function of z.

Thus $\mu_3 = 0 = \mu_5 = \ldots$

The even order moments about the mean are obtained as below:

$$\mu_{2n} = E(X-\mu)^{2n}$$

$$= \int_{-\infty}^{\infty} (x-\mu)^{2n} \cdot \frac{1}{\sigma\sqrt{2\pi}} e^{-\frac{(x-\mu)^2}{2\sigma^2}} dx$$

$$= \int_{-\infty}^{\infty} (z\sigma)^{2n} \cdot \frac{1}{\sigma.\sqrt{2\pi}} e^{\frac{z^2}{2}} \sigma dz, \text{ on putting } z = \frac{x-\mu}{\sigma}$$

$$= \frac{\sigma^{2n}}{\sqrt{2\pi}} \int_{-\infty}^{\infty} z^{2n} . e^{\frac{z^2}{2}} dz,$$

$$= \frac{2\sigma^{2n}}{\sqrt{2\pi}} \int_{0}^{\infty} z^{2n} . e^{\frac{z^2}{2}} dz,$$

Let $y = \frac{z^2}{2}$ Then $dy = z\,dz,$ and therefore,

$$\mu_{2n} = \frac{\sigma^{2n}}{\sqrt{2\pi}} 2^{n+\frac{1}{2}} \int_{0}^{\infty} y^{n-\frac{1}{2}} e^{-y} dt$$

$$=\frac{2^n \sigma^{2n}}{\sqrt{\pi}} \cdot \int_0^\infty y^{\left(n+\frac{1}{2}\right)-1} e^{-y}\, dy$$

$$=\frac{\tilde{A}^n.\sigma^{2n} \left(n+\frac{1}{.2}\right)}{\sqrt{\pi}}.$$

$$=\frac{2^n.\sigma^{2n}}{\sqrt{\pi}} \cdot \left(\frac{2n-1}{2}\right) \cdot \left(\frac{2n-3}{2}\right) \cdots \left(\frac{5}{2}\right)\left(\frac{3}{2}\right)\left(\frac{1}{2}\right)\sqrt{\pi}$$

$$=(2n-1)(2n-3)\ldots 5.3.1\sigma^{2n}$$

Putting $n=1\,and\,2,$ we get $\mu_2=\sigma^2$ and $\mu_4=3\sigma^4$.

Hence $\beta_1=0$ meaning that skewness is zero, and $\beta_2=\frac{\mu_4}{\mu_2^2}=\frac{3\sigma^4}{\sigma^4}=3,$ i.e.

The normal curve has zero kurtosis. This is, in fact, the origin of the choice of the apparently arbitrary value 3 in the definition of the platy and leptokurtosis.

- If X is $N(\mu,\sigma^2)$ and if $Y=a+bX$, then Y is $N(a+b\mu,b^2\sigma^2)$.

Proof. Finding expectation and variance of Y, we get

$E(Y)=E(a+bX)=a+b\mu,(\because a\,and\,b\,are\,constants)$

And $var(Y)=E(Y^2)-[E(Y)]^2=b^2\sigma^2$

Now the function $X=a+bX$ may be either a decreasing or increasing function according as b is negative or positive. If Y is normally distributed, then its p.d.f. is given by

$$h(y)=\frac{1}{\sigma\sqrt{2\pi}}e^{-\frac{\left(\frac{y-a}{b}-\mu\right)^2}{2\sigma^2}}.\left|\frac{1}{b}\right| \quad \{\because h(y)=f(x)\left|\frac{dx}{dy}\right|$$

$$= \frac{1}{|b|\sigma\sqrt{2\pi}} e^{-\frac{[y-(a+b\mu)]^2}{2b^2\sigma^2}}$$

Which represents the p . d . f. of a r.v. that is $N(a+b\mu, b^2\sigma^2)$

It follows that, if $X \, is \, N(\mu, \sigma^2)$ and if $Z = \frac{X-\mu}{\sigma}, Z \, is \, N(0,1)$.

- The sum of independent normal variables is a normal variable. Stated differently, if X_1 is $N(\mu_1, \sigma_1^2)$ and X_2 is $N(\mu_2, \sigma_2^2)$, then for independent $X_1 \, and \, X_2, X_1 + X_2 \, is$

$N(\mu_1 + \mu_2, \sigma_1^2 + \sigma_2^2)$.

- No matter what the values of μ and σ are, areas under normal curve remain in certain fixed proportions within a specified number of standard deviations on either side of μ. For example, the interval

  _$\mu \pm \sigma$ will always contain 68.26%

  _$\mu \pm 2\sigma$ will always contain 95.44%

  _$\mu \pm 3\sigma$ will always contain 99.73%

Practically all of the area is between $\mu - 3\sigma \, and \, \mu + 3\sigma$, the range of the distribution is therefore approximately 6 standard deviations (theoretically curve goes from $-\infty \, to \, \infty$), and we usually terminate the graph at these points. This is a very important property of the normal distribution as most of the tests of significance for large samples are based on it.

- The quartile deviation, Q is found as

$$\frac{1}{\sigma\sqrt{2\pi}} \int_{\mu-Q}^{\mu+Q} e^{-\frac{(x-\mu)^2}{2\sigma^2}} dx = \frac{1}{2}$$

or

$$\frac{1}{\sqrt{2\pi}}\int_{0}^{\frac{Q}{\sigma}} e^{-\frac{z^2}{2}}\,dz = \frac{1}{4}, where\, z = \frac{x-\mu}{\sigma}$$

We find, from area tables that $\frac{Q}{\sigma} = 0.6745\, or\, Q = 0.6745\sigma$

Which is also called the probable error, a term not used nowadays. This also gives the values of the quartiles which are

$Q_1 = \mu - 0.6745\sigma\, and\, Q_3 = \mu + 0.6745\sigma.$

- The normal curve approaches, but never really touches, the horizontal axis on either side of the mean towards plus and minus infinity, that is the curve is asymptotic to the horizontal axis as $x \to \pm\infty$.

***Example*:** For a certain normal distribution, the first moment about 10 is 40 and the fourth moment about about 50 is 48. Find its mean and standard deviation.

***Solution*:** For a normal distribution, the first moment about an arbitrary mean a is given by

$$\mu_1' = \int_{-\infty}^{\infty}(x-a)f(x)\,dx, where\, f(x) = \frac{1}{\sigma\sqrt{2\pi}}e^{-\frac{(x-\mu)^2}{2\sigma^2}}$$

$$40 = \int_{-\infty}^{\infty}(x-10)f(x)\,dx$$

$$40 = \int_{-\infty}^{\infty}xf(x)\,dx - 10\int_{-\infty}^{\infty}f(x)\,dx$$

$40 = \mu - 10,$ where μ denotes mean

$\mu = 50$.

The fourth moments about 50 (which is mean) implies that $\mu_4 = 48.$ But for a normal distribution, the fourth moment about the mean is $3\sigma^4$.

Therefore, we have

$3\sigma^4 = 48 \, which \, gives \, \sigma = 2$

Hence mean = 50 and standard deviation = 2.

Moment Generating and Cumulant Generating Function of the Normal Distribution

Let the r.v. X be $N(\mu, \sigma^2)$. Then the m.g.f of X with respect to origin is given by

$$M_0(t) = E(e^{tX}) = \frac{1}{\sigma\sqrt{2\pi}} \int_{-\infty}^{\infty} e^{tx} . e^{-\frac{(x-\mu)^2}{2\sigma^2}} dx$$

Let $z = \frac{x-\mu}{\sigma}$ so that $dx = \sigma dz$. Then

$$M_0(t) = \frac{1}{\sigma\sqrt{2\pi}} \int_{-\infty}^{\infty} e^{t(\mu + z\sigma)} e^{-\frac{z^2}{2}} \sigma \, dz$$

$$= \frac{e^{\mu t}}{\sqrt{2\pi}} \int_{-\infty}^{\infty} e^{tz\sigma - \frac{z^2}{2}} dz$$

$$= \frac{e^{\mu t}}{\sqrt{2\pi}} \int_{-\infty}^{\infty} e^{-\frac{1}{2}\left(z^2 - 2tz\sigma + t^2\sigma^2 - t^2\sigma^2\right)} dz$$

$$= e^{\mu t + \frac{1}{2}t^2\sigma^2} . \frac{1}{\sqrt{2\pi}} \int_{-\infty}^{\infty} e^{-\frac{1}{2}(z - t\sigma)^2} dz$$

$$= e^{\mu t + \frac{1}{2}t^2 z^2}$$

The m.g.f with respect to mean, μ, is

$$M_\mu(t) = E\left[e^{t(X-\mu)}\right] = e^{-\mu t} E(e^{tX})$$

$$= e^{-\mu t} . e^{\mu t + \frac{1}{2}t^2\sigma^2} = e^{\frac{1}{2}t^2\sigma^2}$$

$$= 1 + \frac{\left(\frac{1}{2}t^2\sigma^2\right)}{1!} + \frac{\left(\frac{1}{2}t^2\sigma^2\right)^2}{2!} + \ldots + \frac{\left(\frac{1}{2}t^2\sigma^2\right)^n}{n1}$$

Hence $\mu_{2n+1} = 0;$ and

$$\mu_{2n} = co - efficient\, of\, \frac{t^{2n}}{(2n)!}$$

$$= 1.3.5\ldots(2n-1)\sigma^{2n}, \quad for\, all\, n \geq 0.$$

The cumulant generating function is given by

$$k(t) = \log_e M_0(t) = \mu t + \frac{1}{2}t^2\sigma^2$$

Thus $k_1 = \mu,\ k_2 = \sigma^2$ and $k_r = 0\ for\ r \geq 3.$

Hence all the cumulant after the second are equal to zero.

Example: If the p.d.f. of the r.v. X is

$$f(x) = \frac{1}{\sqrt{32\pi}} e^{-\frac{(x+7)^2}{32}} \qquad -\infty < x < \infty,$$

Find its mean, variance and moment generating function.

Solution: The function $f(x)$ may be written as

$$f(x) = \frac{1}{4\sqrt{2\pi}} e^{-\frac{[x-(-7)]^2}{2(4)^3}}$$

Comparing with the general form of the normal distribution, we find that $\mu = -7\, and\, \sigma^2 = 16.$

Hence X is $N(-7,16)$.

Substituting the values of mean and variance in the relation m.g.f.

$= e^{\mu t + \frac{1}{2}\sigma^2 t^2}$, we get

$m.g.f. = e^{-7t+\frac{1}{2}(16)t^2} = e^{-7t+8t^2}$

Example: Let $X = N(0,1)$ mean that X has a normal distribution with zero mean and unit variance. What will be the distributions of $2X-3; \frac{7}{8}X+5 \, and \, 4X$?

Solution: Given that $E(X) = 0 \, and \, Var(X) = 1.$

Let $Y = 2X-3$. Then $E(Y) = E(2x-3) = 2E(X)-3 = -3,$ and $Var(y) = Var(2X-3) = 2^2 \, Var(X) = 4$

Hence the distribution of $2X-3 \, is \, N(-3,4).$

Similarly, we find that $\frac{7}{8}X+5 \, is \, N\left(5, \frac{49}{64}\right) and \, 4X \, is \, N(0,16).$

Example: If $X \, is \, N(0,1)$, then find the distribution of X^2.

Solution: The distribution of X is

$$dF = \frac{1}{\sqrt{2\pi}} e^{-\frac{x^2}{2}} dx, -\infty < x < \infty$$

$$= \frac{1}{\sqrt{2\pi}} e^{-\frac{y}{2}} . y^{-\frac{1}{2}} dy$$

$$= \frac{1}{\tilde{A}\left(\frac{1}{2}\right)} e^{-m} . m^{\frac{1}{2}-1} dm, where \, m = \frac{y}{2} and \, 0 \leq m \leq \infty.$$

This is Gamma variable with parameter $\frac{1}{2}$.

Tabulated Area of Normal Distribution

The area under the normal curve between any two ordinates at $X = a$ and $X = b$ equals the probability that the r.v. X lies in the interval $[a,b]$. That is

$$P(a \le X \le b) = \int_a^b \frac{1}{\sigma\sqrt{2\pi}} e^{-\frac{(x-\mu)^2}{2\sigma^2}} dx,$$

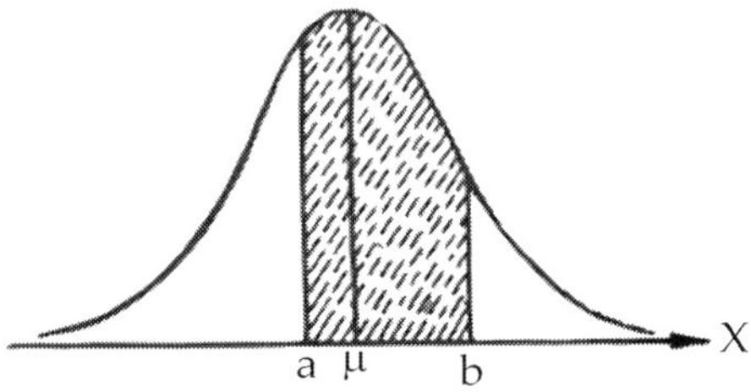

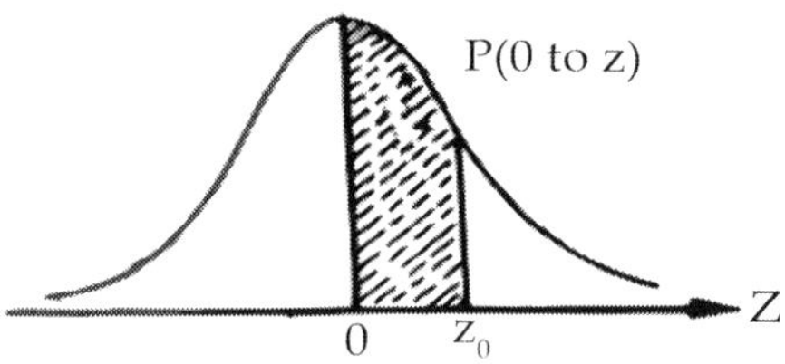

Which is represented by the area of the shaded region.

But integrals of this type cannot be solved by ordinary means. They are, however, evaluated by the methods of numerical integration, and numerical approximations for some functions have been tabulated for quick reference.

From Table the areas for the standard normal distribution from the mean $z = 0$ to a specified positive value of $z, say\ z_0$. Since normal curves are symmetrical, therefore $P(0\,to\,z) = P(0\,to - z)$

That is why the areas for negative values of z are not tabulated. It is important to note that this single table for the standard normal distribution suffices for the calculation of probabilities for any normal distribution. The values of the r.v. X in any problem are changed to the values of the standard normal variable Z and the desired probabilities are obtained from table. Thus to find $P(a < X < b)$, we would change X into Z as follows:

$$P(a \leq X \leq b) = P\left(\frac{a-\mu}{\sigma} \leq \frac{X-\mu}{\sigma} \leq \frac{b-\mu}{\sigma}\right)$$

$$= P\left(\frac{a-\mu}{\sigma} \leq Z \leq \frac{b-\mu}{\sigma}\right)$$

Where $\frac{a-\mu}{\sigma}$ and $\frac{b-\mu}{\sigma}$ are the z-values of the standard normal variable Z. In practice, a normal curve sketch for the given problem, showing under the $X-$ scale for the corresponding values of z will help in solving the problem.

Table of the standard normal distribution values (*z* £ 0)

−z	0.00	0.01	0.02	0.03	0.04	0.05	0.06	0.07	0.08	0.09
0.0	0.50000	0.49601	0.49202	0.48803	0.48405	0.48006	0.47608	0.47210	0.46812	0.46414
0.1	0.46017	0.45621	0.45224	0.44828	0.44433	0.44038	0.43644	0.43251	0.42858	0.42466
0.2	0.42074	0.41683	0.41294	0.40905	0.40517	0.40129	0.39743	0.39358	0.38974	0.38591
0.3	0.38209	0.37828	0.37448	0.37070	0.36693	0.36317	0.35942	0.35569	0.35197	0.34827
0.4	0.34458	0.34090	0.33724	0.33360	0.32997	0.32636	0.32276	0.31918	0.31561	0.31207
0.5	0.30854	0.30503	0.30153	0.29806	0.29460	0.29116	0.28774	0.28434	0.28096	0.27760
0.6	0.27425	0.27093	0.26763	0.26435	0.26109	0.25785	0.25463	0.25143	0.24825	0.24510
0.7	0.24196	0.23885	0.23576	0.23270	0.22965	0.22663	0.22363	0.22065	0.21770	0.21476
0.8	0.21186	0.20897	0.20611	0.20327	0.20045	0.19766	0.19489	0.19215	0.18943	0.18673
0.9	0.18406	0.18141	0.17879	0.17619	0.17361	0.17106	0.16853	0.16602	0.16354	0.16109
1.0	0.15866	0.15625	0.15386	0.15151	0.14917	0.14686	0.14457	0.14231	0.14007	0.13786
1.1	0.13567	0.13350	0.13136	0.12924	0.12714	0.12507	0.12302	0.12100	0.11900	0.11702
1.2	0.11507	0.11314	0.11123	0.10935	0.10749	0.10565	0.10384	0.10204	0.10027	0.09853
1.3	0.09680	0.09510	0.09342	0.09176	0.09012	0.08851	0.08692	0.08534	0.08379	0.08226
1.4	0.08076	0.07927	0.07780	0.07636	0.07493	0.07353	0.07215	0.07078	0.06944	0.06811
1.5	0.06681	0.06552	0.06426	0.06301	0.06178	0.06057	0.05938	0.05821	0.05705	0.05592
1.6	0.05480	0.05370	0.05262	0.05155	0.05050	0.04947	0.04846	0.04746	0.04648	0.04551
1.7	0.04457	0.04363	0.04272	0.04182	0.04093	0.04006	0.03920	0.03836	0.03754	0.03673
1.8	0.03593	0.03515	0.03438	0.03363	0.03288	0.03216	0.03144	0.03074	0.03005	0.02938
1.9	0.02872	0.02807	0.02743	0.02680	0.02619	0.02559	0.02500	0.02442	0.02385	0.02330
2.0	0.02275	0.02222	0.02169	0.02118	0.02068	0.02018	0.01970	0.01923	0.01876	0.01831
2.1	0.01786	0.01743	0.01700	0.01659	0.01618	0.01578	0.01539	0.01500	0.01463	0.01426

2.2	0.01390	0.01355	0.01321	0.01287	0.01255	0.01222	0.01191	0.01160	0.01130	0.01101
2.3	0.01072	0.01044	0.01017	0.00990	0.00964	0.00939	0.00914	0.00889	0.00866	0.00842
2.4	0.00820	0.00798	0.00776	0.00755	0.00734	0.00714	0.00695	0.00676	0.00657	0.00639
2.5	0.00621	0.00604	0.00587	0.00570	0.00554	0.00539	0.00523	0.00509	0.00494	0.00480
2.6	0.00466	0.00453	0.00440	0.00427	0.00415	0.00403	0.00391	0.00379	0.00368	0.00357
2.7	0.00347	0.00336	0.00326	0.00317	0.00307	0.00298	0.00289	0.00280	0.00272	0.00264
2.8	0.00256	0.00248	0.00240	0.00233	0.00226	0.00219	0.00212	0.00205	0.00199	0.00193
2.9	0.00187	0.00181	0.00175	0.00170	0.00164	0.00159	0.00154	0.00149	0.00144	0.00140
3.0	0.00135	0.00131	0.00126	0.00122	0.00118	0.00114	0.00111	0.00107	0.00104	0.00100
3.1	0.00097	0.00094	0.00090	0.00087	0.00085	0.00082	0.00079	0.00076	0.00074	0.00071
3.2	0.00069	0.00066	0.00064	0.00062	0.00060	0.00058	0.00056	0.00054	0.00052	0.00050
3.3	0.00048	0.00047	0.00045	0.00043	0.00042	0.00040	0.00039	0.00038	0.00036	0.00035
3.4	0.00034	0.00033	0.00031	0.00030	0.00029	0.00028	0.00027	0.00026	0.00025	0.00024
3.5	0.00023	0.00022	0.00022	0.00021	0.00020	0.00019	0.00019	0.00018	0.00017	0.00017

Table of the standard normal distribution values (*z* ³ 0)

z	**0.00**	**0.01**	**0.02**	**0.03**	**0.04**	**0.05**	**0.06**	**0.07**	**0.08**	**0.09**
0.0	0.50000	0.50399	0.50798	0.51197	0.51595	0.51994	0.52392	0.52790	0.53188	0.53586
0.1	0.53983	0.54380	0.54776	0.55172	0.55567	0.55962	0.56356	0.56749	0.57142	0.57535
0.2	0.57926	0.58317	0.58706	0.59095	0.59483	0.59871	0.60257	0.60642	0.61026	0.61409
0.3	0.61791	0.62172	0.62552	0.62930	0.63307	0.63683	0.64058	0.64431	0.64803	0.65173
0.4	0.65542	0.65910	0.66276	0.66640	0.67003	0.67364	0.67724	0.68082	0.68439	0.68793
0.5	0.69146	0.69497	0.69847	0.70194	0.70540	0.70884	0.71226	0.71566	0.71904	0.72240
0.6	0.72575	0.72907	0.73237	0.73565	0.73891	0.74215	0.74537	0.74857	0.75175	0.75490
0.7	0.75804	0.76115	0.76424	0.76730	0.77035	0.77337	0.77637	0.77935	0.78230	0.78524
0.8	0.78814	0.79103	0.79389	0.79673	0.79955	0.80234	0.80511	0.80785	0.81057	0.81327
0.9	0.81594	0.81859	0.82121	0.82381	0.82639	0.82894	0.83147	0.83398	0.83646	0.83891
1.0	0.84134	0.84375	0.84614	0.84849	0.85083	0.85314	0.85543	0.85769	0.85993	0.86214
1.1	0.86433	0.86650	0.86864	0.87076	0.87286	0.87493	0.87698	0.87900	0.88100	0.88298
1.2	0.88493	0.88686	0.88877	0.89065	0.89251	0.89435	0.89617	0.89796	0.89973	0.90147
1.3	0.90320	0.90490	0.90658	0.90824	0.90988	0.91149	0.91308	0.91466	0.91621	0.91774
1.4	0.91924	0.92073	0.92220	0.92364	0.92507	0.92647	0.92785	0.92922	0.93056	0.93189
1.5	0.93319	0.93448	0.93574	0.93699	0.93822	0.93943	0.94062	0.94179	0.94295	0.94408

1.6	0.94520	0.94630	0.94738	0.94845	0.94950	0.95053	0.95154	0.95254	0.95352	0.95449
1.7	0.95543	0.95637	0.95728	0.95818	0.95907	0.95994	0.96080	0.96164	0.96246	0.96327
1.8	0.96407	0.96485	0.96562	0.96638	0.96712	0.96784	0.96856	0.96926	0.96995	0.97062
1.9	0.97128	0.97193	0.97257	0.97320	0.97381	0.97441	0.97500	0.97558	0.97615	0.97670
2.0	0.97725	0.97778	0.97831	0.97882	0.97932	0.97982	0.98030	0.98077	0.98124	0.98169
2.1	0.98214	0.98257	0.98300	0.98341	0.98382	0.98422	0.98461	0.98500	0.98537	0.98574
2.2	0.98610	0.98645	0.98679	0.98713	0.98745	0.98778	0.98809	0.98840	0.98870	0.98899
2.3	0.98928	0.98956	0.98983	0.99010	0.99036	0.99061	0.99086	0.99111	0.99134	0.99158
2.4	0.99180	0.99202	0.99224	0.99245	0.99266	0.99286	0.99305	0.99324	0.99343	0.99361
2.5	0.99379	0.99396	0.99413	0.99430	0.99446	0.99461	0.99477	0.99492	0.99506	0.99520
2.6	0.99534	0.99547	0.99560	0.99573	0.99585	0.99598	0.99609	0.99621	0.99632	0.99643
2.7	0.99653	0.99664	0.99674	0.99683	0.99693	0.99702	0.99711	0.99720	0.99728	0.99736
2.8	0.99744	0.99752	0.99760	0.99767	0.99774	0.99781	0.99788	0.99795	0.99801	0.99807
2.9	0.99813	0.99819	0.99825	0.99831	0.99836	0.99841	0.99846	0.99851	0.99856	0.99861
3.0	0.99865	0.99869	0.99874	0.99878	0.99882	0.99886	0.99889	0.99893	0.99896	0.99900
3.1	0.99903	0.99906	0.99910	0.99913	0.99916	0.99918	0.99921	0.99924	0.99926	0.99929
3.2	0.99931	0.99934	0.99936	0.99938	0.99940	0.99942	0.99944	0.99946	0.99948	0.99950
3.3	0.99952	0.99953	0.99955	0.99957	0.99958	0.99960	0.99961	0.99962	0.99964	0.99965
3.4	0.99966	0.99968	0.99969	0.99970	0.99971	0.99972	0.99973	0.99974	0.99975	0.99976
3.5	0.99977	0.99978	0.99978	0.99979	0.99980	0.99981	0.99981	0.99982	0.99983	0.99983

***Example*:** Let the r.v. have the standard normal distribution. Find

$i.P(0\le Z\le 1.20)$, $ii.P(-1.65\le Z\le 0)$ $iii.P(0.6\le Z\le 1.67)$

$iv.P(-1.30\le Z\le 2.18)v.P(-1.96\le Z\le -0.84)vi.P(Z\ge 1.96)vi.P(Z\le -2.15)$

***Solution*:** First we draw the normal curve sketch, shading the desired area for each part.

- To find $P(0\le Z\le 1.20)$, from the table, we move downward the column marked Z until 1.2 is reached, and then move across that row to the column headed 0.00 to find entry 0.3849.

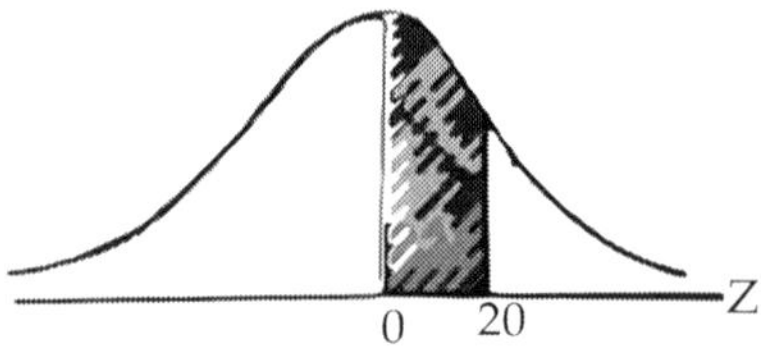

Therefore $P(0 \le Z \le 1.20) = 0.3849$

- Since the normal curve is symmetrical about the mean, therefore area between $z = 0$ and a positive value of z is equal to the area between $z = 0$ and a negative value of z of the same magnitude.

Hence using table, we can solve

$P(-1.65 \le Z \le 0) = P(0 \le Z \le 1.65) = 0.4505$

$P(0.6 \le Z \le 1.67)$

$= P(0 \le Z \le 1.67) - P(0 \le z \le 0.6)$

$= 0.4525 - 0.2257$

$= 0.2268$

$P(-1.30 < Z \le 2.18)$

$= P(-1.30 \le Z \le 0) + P(0 \le Z \le 2.18)$

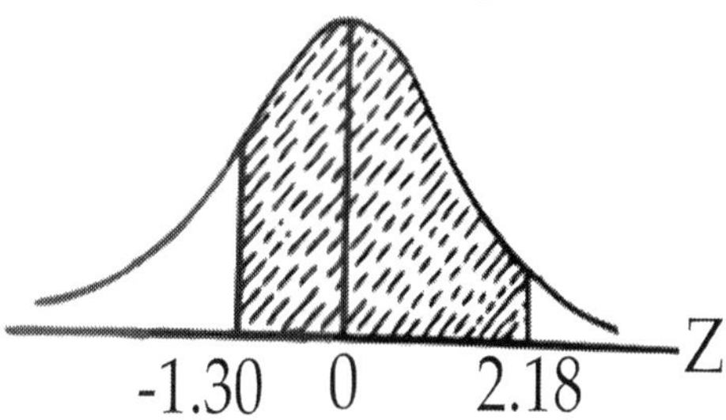

$= 0.4032 + 0.4854$

$= 0.8886(\textit{from area table})$

$P(-1.96 \le Z \le -0.84)$

$= P(-1.96 \leq Z \leq 0) - P(-0.84 \leq Z \leq 0)$

$= 0.4750 - 0.2995$

$= 0.1755(from\, area\, table)$

$P(Z \geq 1.96)$

$= 0.5 - P(0 \leq Z \leq 1.96)$

$= 0.5 - 0.475$

$= 0.0250$

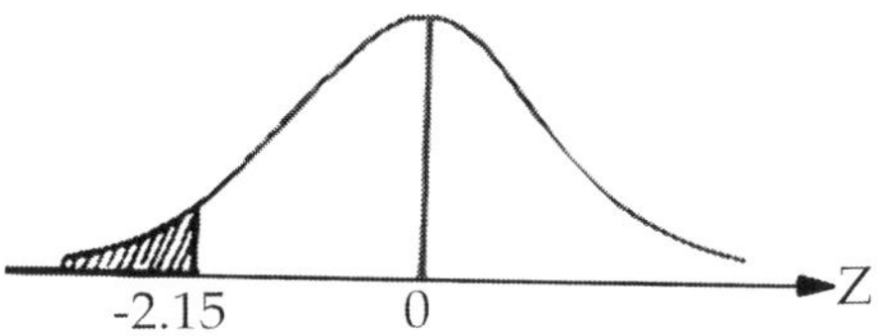

$P(Z \leq -2.15)$

$= 0.5 - 0.4842$

$= 0.0158$

***Example*:** A random variable X is normally distributed with $\mu = 50\, and\, \sigma^2 = 25.$ Find the probability (a) that it will fall between i) 0 and 40, ii) 55 and 100; (b) that it will be i) larger than 54, ii) smaller than 57.

***Solution*:** We draw the normal curve sketch showing $x\, and\, z$ values, and the desired area for each part. With $\mu = 50\, and\, \sigma = 5,$ we have

$$z = \frac{x - \mu}{5}$$

(a)

- At $x = 0$, we compute $z = \frac{0 - 50}{5} = -10,$ and at $x = 40,$ we find $z = \frac{40 - 50}{5} = -2$

So $P(0 \le X \le 40)$

$= P(-10 \le Z \le -2)$

$= P(-10 \le Z \le 0) - P(-2 \le Z \le 0)$

$= 0.5 - 0.4772$

$= 0.0228$

- We have for $x = 55,$

$z = \dfrac{55-50}{5} = 1$

For $x = 100, z = \dfrac{100-50}{5} = 10$

The x values and the corresponding z values are shown in figure

$P(55 \le X \le 100) = P(1 \le Z \le 10)$

$= P(0 \le Z \le 10) - P(0 \le Z \le 1)$

$= 0.5 - 0.3413 = 0.1587$

(b)

i) With $\mu = 55\, and\, \sigma = 5,$ we have

for $x = 54, z = \dfrac{54-50}{5} = 0.80$

$P(X \ge 54) = P(Z \ge 0.8)$

$= 0.5 - P(0 \le Z \le 0.8)$

$= 0.5 - 0.2881$

$= 0.2119$

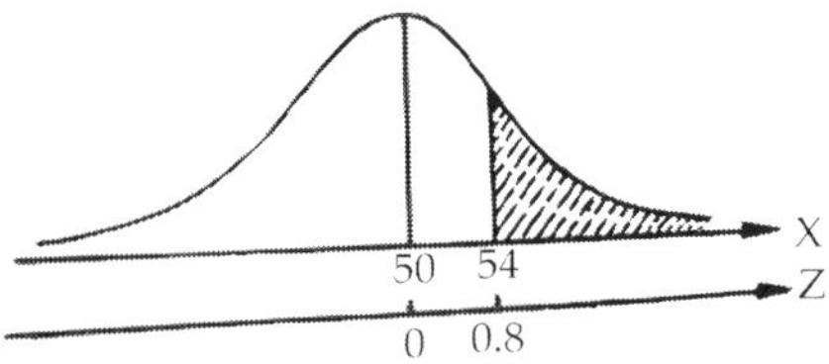

ii) At $x = 57, z = \dfrac{57-50}{5} = 1.40$

$P(X < 57) = P(Z < 1.40)$

$= 0.5 + P(0 \le Z \le 1.40)$

$= 0.5 + 0.4192$

$= 0.9192$

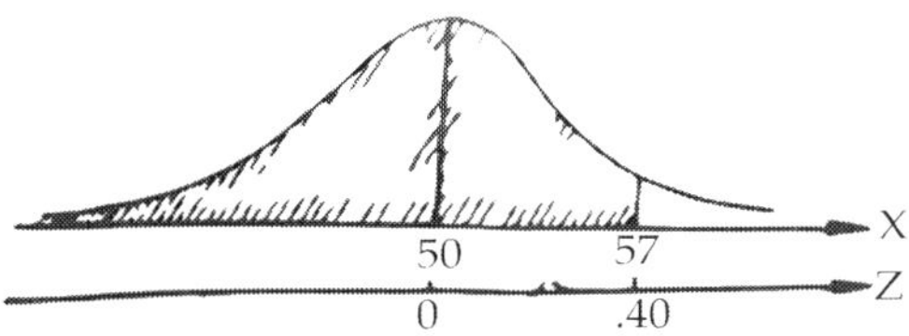

Example: The length of life for an automatic dishwasher is approximately normally distributed, with a mean of 3.5 years and a standard deviation of 1.0 years. If this type of dishwasher is guaranteed for 12 months, what fraction of the sales will require replacement?

Solution: The fraction of sales requiring replacement is equal to the area under the normal curve for $X \le 1$ year, the guaranteed period.

With $\mu = 3.5\, and\, \sigma = 1.0$, we therefore transform $x = 1.0$ year to the corresponding z value. Thus we find

$$z = \frac{1.0 - 3.5}{1.0} = -2.5$$

The x values and the corresponding z values are indicated in the figure. From Table

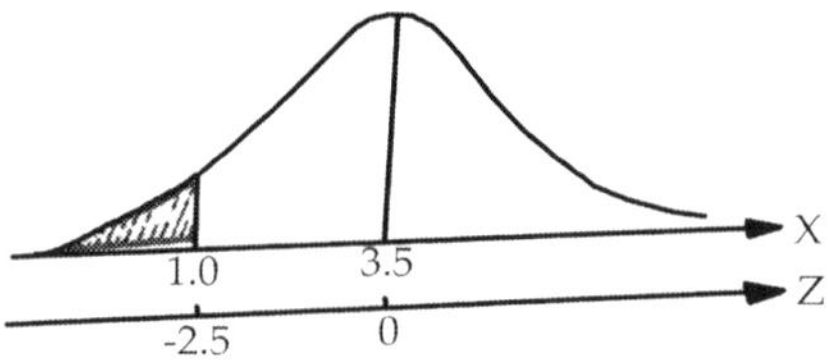

$P(X \leq 1.0) = P(Z \leq -2.5)$

$= 0.5 - P(-2.5 \leq Z \leq 0)$

$= 0.5 - 0.4938 = 0.0062.$

Hence $0.62\ \%$ of sales need replacement before 12 months.

Example: The mean height of soldiers is 68.22 inches with a variance of 0.8 $(in)^2$. Assuming the distribution of heights to be normal, how many soldiers in a regiment of 1000 would you expect to be over 6 feet tall?

Solution: With $\mu = 68.22$ and $\sigma^2 = 10.8(in.)^2$, we first compute the z value.

At $x = 72(6\ ft.)$, we have

$$z = \frac{72 - 68.22}{\sqrt{10.8}} = 1.15$$

The x values and the corresponding z values are shown in normal curve sketch, and we need the right tail area which is shaded.

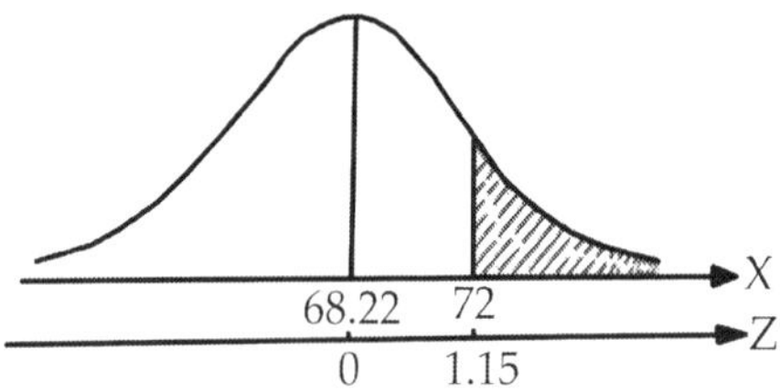

Therefore using table, we find

$P(X \geq 72) = P(Z \geq 1.15) = 0.5 - P(0 \leq Z \leq 1.15)$

$= 0.5 - 0.3749$

$= 0.1251$

If there are $1,000$ soldiers in the regiment, then number expected to be over 6 feet (or 72 inches) is
$1,000 \times 0.1251 = 125.$

Example: If the moment generating function of X is $M(t) = e^{166\,t + 200\,t^2}$, find

(i) $P(170 < X < 200)$ (ii) $P(148 \leq X \leq 172).$

Solution: Comparing $M(t) = e^{166\,t + 200t^2}$ with the m.g.f of $N(\mu, \sigma^2)$, we find $\mu = 166\, and\, \sigma^2 = 400.$

To find the desired probabilities, we transform x values to z values, using $z = \frac{X - 166}{20}.$ Therefore

- at $x = 170,$ we get $z = \frac{170 - 166}{20} = 0.2,$ and at $x = 200,$ we find $z = \frac{200 - 166}{20} = 1.7.$

Hence using table, we find

$P(170 < X < 200) = P(0.2 < Z < 1.7 = P(0 < Z < 1.7) - P(0 < Z < 0.2)$

$= 0.4554 - 0.0793 = 0.3761$

$P(148 \leq X \leq 170) = P(-0.9 < Z < 0.3)$

$= P(-0.9 \leq Z \leq 0) + P(0 \leq Z \leq 0.3)$

$= 0.3159 + 0.1179 = 0.4338$

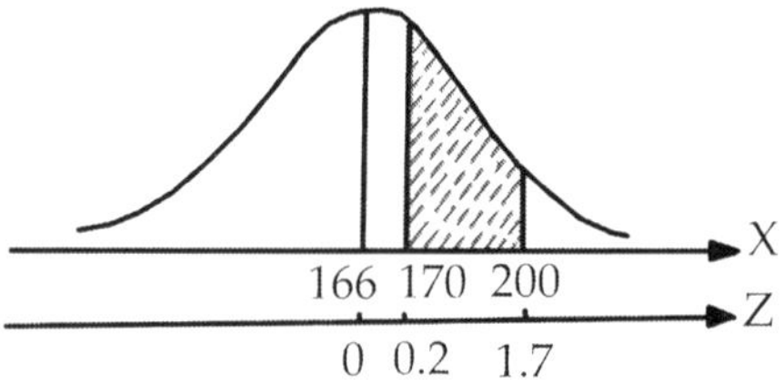

Inverse Use of Table of Areas under Normal Curve

When we use the table of areas under the normal curve to find the value of z corresponding to a given probability in the body of the table, it is called the inverse use of the area table. The value of z given that $P(0 \le Z \le z_0) = k$ may be denoted by the symbol $(z|P = k)$. If the given probability does not appear in the table, the closest probability may be taken to find the corresponding value of z. The z value thus obtained is then changed to an x value, using the formula

$$z = \frac{x - \mu}{\sigma}$$

Which may be written as $x = \mu + z\sigma$.

The value of z is positive when x lies to the right of μ and it is negative when x lies to the left of μ.

A sketch of the standard normal curve showing the given probability and the location of x on the correct side of μ helps in solving the given problem.

Example: In a normal distribution $\mu = 40$ and $P(25 \le X \le 55) = 0.8662.$ Find $P(20 \le X \le 60)$.

Solution: Given $\mu = 40$ and $P(25 \le X \le 55) = 0.8662,$ where X is a normal random variable. We find that 25 and 55 lie on either side of the mean $\mu = 40$ at equal distance, so that

$$P(between\, X = 25\, and\, \mu = 40) = P(\mu = 40\, and\, X = 55) = 0.4331$$

Therefore, using Table inversely, we find

$z \mid P = 0.4331 = 1.50$

Substitution in the relation $z = \frac{x-\mu}{\sigma}$ gives

$1.50 = \frac{55-40}{\sigma}$ which yields $\sigma = 10$

Now $P(20 \le X \le 60) = p\left(\frac{20-40}{10} \le \frac{X-40}{10} \le \frac{60-40}{10}\right) = P(-2 \le Z \le 2)$

$= 0.4772 + 0.4772 = 0.9544$

Example: The time required by a nurse to inject a shot of penicillin has been observed to be normally distributed, with a mean of $\mu = 30$ seconds and a standard deviation of $\sigma = 10$ seconds. Find the following percentiles: (i) 10th, (ii) 90th.

- Let x be the point P_{10}, where P_{10} is a point at or below which 10% of the area lies. Then the area to the left of x is 0.1 and area between μ and x is $0.5 - 0.1 = 0.4$

Looking at table we find that a probability of 0.4 does not appear in the table, so we take the closest probability to 0.4, which is 0.3997.

Thus, using table inversely we find $\left(\frac{z}{P} = 0.3997\right) = 1.28$

Since x lies to the left of μ, therefore z is negative at this point. Hence $x = \mu + z\sigma$

$= 30 + (-1.28)(10) = 17.2\,seconds.$

- Let x be the point P_{90}, where P_{90} is a point at or below which 90% of the area lies. Then the area to the left of x is 0.9 and the area between μ and x is 0.9 – 0.5 = 0.4.

From table, we find $\left(\frac{z}{P} = 0.4\right) = 1.28,$ Since x lies to the right of μ, therefore z is positive at the point.

Hence $x = \mu + z\sigma = 30 + (1.28)(10) = 42.8\,seconds$

***Example*:** In a normal distribution with $\mu = 13.5\, and\, \sigma = 3.6,$ find two points such that a single observation has 95% chance for falling between them.

***Solution*:** Let $x_1\, and\, x_2$ be the two points between which the probability of an observation falling is 0.95. As the curve is symmetrical, so half of 0.95, i.e. 0.475 is the area lying on either side of $\mu.$

Now using table inversely, we find that

$$\left(\frac{z}{P} = 0.475\right) = 1.96$$

The point x_1 is to the left of $\mu,$ so $z = -1.96$ and at x_2 which is to the right of $\mu, z = 1.96.$

Therefore,

$x_1 = \mu + z\sigma = 13.5 + (-1.96)(3.6) = 6.4,$ and

$x_2 = \mu + z\sigma = 13.5 + (1.96)(3.6) = 20.6$

***Example*:** An athlete finds that in a high jump he can clear a height of $1.68\,m$ once in five attempts and a height of 1.52 m nine times out of ten attempts. Assuming the heights he can clear in various jumps form a normal distribution, estimate the mean and standard deviation of the distribution.

***Solution*:** Let X denote the height the athlete can clear in various jumps. Then X is $N(\mu, \sigma^2)$ where $\mu\, and\, \sigma^2$ are unknown.

Now there is a probability of $\frac{1}{5} = 0.2,$ if he can clear a height of $1.68\,m, i.e.$

$P(X > 1.68) = 0.2$ and there is a probability of $\frac{9}{10} = 0.9,$ if he can clear a height of 1.52m, that is $P(X \geq 1.52) = 0.9,$ implying that $P(X \leq 1.52) = 0.1.$

The probability between μ and $x = 1.68$ is $0.5 - 0.2 = 0.3,$ and between $\mu \, and \, x = 1.52$ is $0.5 - 0.1 = 0.4.$

Using Table inversely, we find that $\left(\frac{z}{P} = 0.3\right) = 0.84.$

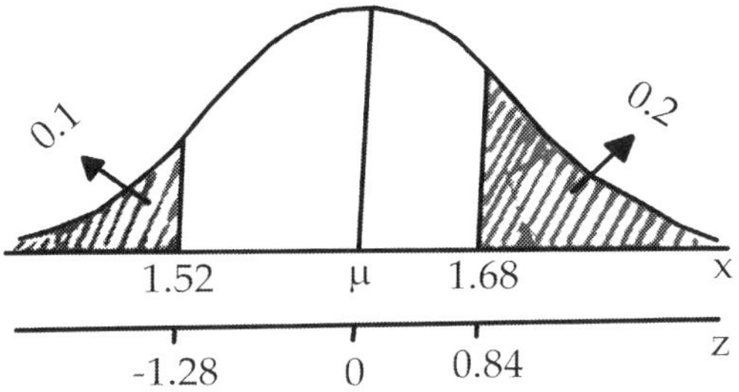

Example: A lawyer commutes daily from his suburban home to his midtown office. On the average, the trip one way takes 24 minutes, with a standard deviation of 3.8 minutes. Assume the distribution of trip times to be normally distributed.

- What is the probability that a trip will take at least $\frac{1}{2}$ hours?
- If the office opens at 9:00 AM and he leaves his house at 8:45 AM daily, what percentage of the time is he left for work?
- If he leaves the house at 8:35 AM and coffee is served at the office from 8:50 AM until 9:00 AM, what is the probability that he misses coffee?
- Find the length of time above which we find the slowest 15 % of the trips.
- Find the probability that 2 of the next 3 trips will take at least $\frac{1}{2}$ hour.

Solution: Let the r.v. X denote the trip time in minutes. Then X is

$N\left[24,(3.8)^2\right]$.

- We need to calculate the probability that a trip will take at least $\frac{1}{2}$ hours, i.e. $P(X \geq 30)$. $\therefore\ P(X \geq 30) = P\left(\frac{X-24}{3.8} \geq \frac{30-24}{3.8}\right)$

$= P(Z \geq 1.58) = 0.0571$

- He leaves home at 8:45 AM and the office opens at 9:00 AM implies that he has 15 minutes to reach the office. He will be late for work if he takes more than 15 minutes. Thus we need $P(X > 15)$.

$$P(X \geq 15) = P\left(\frac{X-24}{3.8} > \frac{15-24}{3.8}\right)$$

$= P(Z > -2.37)$

$= 0.5 + P(0 \leq Z \leq 2.37)$

$= 0.5 + 0.4911 = 0.9911$

Therefore the percentage of the time he will be late for work is 99.11 %.

- He leaves home at 8:35 AM and coffee is served from 8:50AM until 9:00 AM. He will miss coffee if he reaches office after 9:00 AM, i.e. if he takes 25 minutes or more time. Thus we need $P(X \geq 25)$.

$$P(X \geq 25) = P\left(\frac{X-24}{3.8} > \frac{25-24}{3.8}\right)$$

$= P(Z \geq 0.26)$

$= 0.5 - P(0 \leq Z \leq 0.26)$

$= 0.5 - 0.1026 = 0.3974$

- To calculate the length of time (x) above which we find the slowest 15% of the trips, we need to calculate the value of x such that $P(X \geq x) = 0.15 \, or \, P\left(\frac{X-24}{3.8} \geq \frac{x-24}{3.8}\right) = 0.15$

or $P\left(Z \geq \frac{x-24}{3.8}\right) = 0.15$

We find from the tables of standard normal distribution that the value of z corresponding to a tail of 0.15 is 1.04.

$x = \mu + z\sigma = 24 + (1.04)(3.8) = 27.952 \, minutes$

- Using the binomial distribution with $n = 3 \, and \, p = 0.0571$ (where p is the probability that a trip will take at least $\frac{1}{2}$ hour), we find that the probability that 2 out of the next 3 trips will take at least $\frac{1}{2}$ hours is $\binom{3}{2}(0.0571)^2(0.9429) = 0.0092.$

Normal Approximation to the Binomial Distribution.

The binomial distribution $b(x;n,p)$ can be closely approximated by the normal distribution when n is sufficiently large and neither $p \, nor \, q$ is close to zero. As a rule of thumb, the normal distribution provides a reasonable approximation to the binomial distribution if both $np \, and \, nq$ are equal to or greater than 5.

The probability for a binomial random variable X to take the value x is Z.

$$f(x) = \binom{n}{x} p^x q^{n-x}, for \, 0 \leq x \leq n \, and \, q + p = 1$$

The variable X has a mean of $\mu = E(X) = np \, and \, Var(X) = npq.$

To derive the limiting form, we define a new random variable Z by the relation

$$Z = \frac{X - np}{\sqrt{npp}}$$

Now, obviously $E(Z) = 0\, and\, Var(Z) = 1.$

Thus we have to show that Z is normally distributed in the limit as $n \to \infty$ and p is finite. (This is actually the de Moivre-Laplace theorem). We shall use the moment generating function (m. g. f) to prove the theorem.

The m.g.f. of Z is

$$M(t) = E\left(e^{tZ}\right) = E\left[e^{\frac{t(x-np)}{\sqrt{npq}}}\right]$$

$$= e^{-\frac{npt}{\sqrt{npq}}} E\left[e^{\frac{xt}{\sqrt{npq}}}\right]$$

$$= e^{-\frac{npt}{\sqrt{npq}}} \sum_{x=0}^{n} \binom{n}{x} p^x q^{n-x} e^{\frac{xt}{\sqrt{npq}}}$$

$$= e^{-\frac{npt}{\sqrt{npq}}} \sum_{x=0}^{n} \binom{n}{x} \left(pe^{\frac{t}{\sqrt{npq}}}\right)^x q^{n-x}$$

$$= e^{-\frac{npt}{\sqrt{npq}}} \left(q + pe^{\frac{t}{\sqrt{npq}}}\right)^n$$

$$= \left(qe^{-\frac{pt}{\sqrt{npq}}} + pe^{\frac{qt}{\sqrt{npq}}}\right)^n$$

$$= \left[q \frac{\sum_{r=0}^{\infty} \left(\frac{-pt}{\sqrt{npq}}\right)^r}{r!} + p \frac{\sum_{r=0}^{\infty} \left(\frac{qt}{\sqrt{npq}}\right)^r}{r!}\right]^n$$

$$= q\left\{1 - \frac{pt}{\sqrt{npq}} + \frac{1}{2!}\left(\frac{pt}{\sqrt{npq}}\right)^2 - \frac{1}{3!}\left(\frac{pt}{\sqrt{npq}}\right)^3 + \frac{1}{4!}\left(\frac{pt}{\sqrt{npq}}\right)^4 - \ldots\right\}$$

$$+ p\left\{1 + \frac{qt}{\sqrt{npq}} + \frac{1}{2!}\left(\frac{qt}{\sqrt{npq}}\right)^2 + \frac{1}{3!}\left(\frac{qt}{\sqrt{npq}}\right)^3 + \frac{1}{4!}\left(\frac{qt}{\sqrt{npq}}\right)^4 + \ldots\right\}$$

$$= \left[1 + \frac{t^2}{2!}.\frac{1}{n} + \frac{t^3}{3!}.\frac{q-p}{n\sqrt{npq}} + \frac{t^4}{4!}.\frac{q^2 - qp + p^2}{n^2 qp} + \ldots\right]^n$$

Taking logs to the base e, we have

$$\log_e M(t) = n\log_e\left[1 + \frac{t^2}{2!}.\frac{1}{n} + \frac{t^3}{3!}.\frac{q-p}{n\sqrt{npq}} + \frac{t^4}{4!}.\frac{1-3pq}{n^2 qp} + \ldots\right]$$

$$= \frac{t^2}{2!} + \frac{t^3}{3!}.\frac{q-p}{\sqrt{npq}} + \frac{t^4}{4!}.\frac{1-6pq}{nqp} + \ldots$$

Thus $\lim\limits_{n\to\infty} \log_e M(t) = \frac{t^2}{2!}$ or $\lim\limits_{n\to\infty} M(t) = e^{\frac{t^2}{2}}$

We know that $e^{\frac{t^2}{2}}$ is the m.g.f of a standard normal variable. Hence we find that in the limit Z has a standard normal distribution.

It is important to note that a binomial variable is discrete whereas the normal curve probability is a probability for an interval. Therefore in using normal curve areas to approximate binomial probabilities, a discrete value of the binomial variable is to be replaced by an interval before the z values are computed. Accordingly, a discrete value becomes the interval from and this sort of adjustment is called the continuity correction. Thus, the discrete value 5, adjusted means 4.5 to 5.5.

Example: A fair coin is tossed 20 times. Find the probability that the number of heads occurring is between 10 and 14 inclusive by using a) the binomial distribution, b) the normal approximation to the binomial distribution.

Solution: a. Let X denote the number of heads occurring. Then the $p.d$ of X is

$$P(X=x)=f(x)=\binom{20}{x}\left(\frac{1}{2}\right)^{x}\left(\frac{1}{2}\right)^{20-x}$$

and the desired probability is $P(10\leq X\leq 14)$.

Now $$P(10\leq X\leq 14)=\sum_{x=10}^{14}\binom{20}{x}\left(\frac{1}{2}\right)^{20}$$

$$=\binom{20}{10}\left(\frac{1}{2}\right)^{20}+\binom{20}{11}\left(\frac{1}{2}\right)^{20}+\binom{20}{12}\left(\frac{1}{2}\right)^{20}+\binom{20}{13}\left(\frac{1}{2}\right)^{20}+\binom{20}{14}\left(\frac{1}{2}\right)^{20}$$

$$=0.1762+0.1602+0.1201+0.0739+0.0370$$

$$=0.5674$$

Alternatively, $$P(10\leq X\leq 14)=\sum_{x=10}^{14}b(x;20,0.5)$$

$$=\sum_{x=0}^{14}b(x\,;20,0.5)-\sum_{x=0}^{9}b(x\,;20,0.5)$$

$$=0.9793-0.4119$$

$$=0.5674$$

b. Since $np=20(0.5)=10>5\, and\ nq=20(0.5)=10\geq 5,$ so we will use the normal distribution to approximate the binomial distribution

Now $\mu=np=20(0.5)=10, and$

$$\sigma=\sqrt{npq}=\sqrt{20(0.5)(0.5)}=2.24.$$

For the normal approximation, the interval of discrete value $10\leq X\leq 14$ is replaced by the interval $9.5\leq X\leq 14.5$. We compute the z-values as below:

At $x = 9.5$, we find $z = \frac{9.5 - 10}{2.24} = -0.22$, and

at $x = 14.5$, we get $z = \frac{14.5 - 10}{2.24} = 2.01$

Hence, using table, we find

$P(10 \leq X \leq 14) = P(-0.22 \leq Z \leq 2.01)$

$= P(-0.22 \leq Z \leq 0) + P(0 \leq Z \leq 2.01)$

$= 0.0871 + 0.4778$

$= 0.5649$

Hence the probability of obtaining heads between 10 and 14 is 0.5649.

Example: A pair of dice is rolled 180 times. Use the normal approximation method to find the probability that a total of 7 occurs (i) at least 25 times, (ii) between 33 and 41 times inclusive and (iii) exactly 30 times.

Solution: Let X denote the number of times a total of 7 occurs when a pair of dice is rolled. Then X is a binomial variable with $n = 180$ and $p = \frac{1}{6}$ (probability of getting a total of 7 with 2 dice).

Since n is large and p is not too small, so we use the normal approximation method with $\mu = np = 180 \times \frac{1}{6} = 30$, and $\sigma = \sqrt{npq} = \sqrt{180 \times \frac{1}{6} \times \frac{5}{6}} = 5$;

To find the desired probabilities.

- The interval at least 25 includes 25; therefore it starts at 24.5 and extends to ∞, *i.e.* at least 25 becomes the interval 24.5 to ∞.

The corresponding z value is $z = \frac{24.5 - 30}{5} = -1.1$

Hence, using table, we find

$$P(at\,least\,25) = \sum_{x=25}^{180} b\left(x;180,\frac{1}{6}\right)$$

$$= P(-1.1 \leq Z \leq 0) + P(0 \leq Z \leq \infty)$$

$$= 0.3643 + 0.5$$

$$= 0.8643$$

Thus the probability of obtaining at least 25 times seven is 0.8643.

- The interval of discrete values $33 \leq X \leq 41$ is replaced by the interval $32.5 \leq X \leq 41.5,$ and the corresponding z values are

  at $x = 32.5,\ z = \dfrac{32.5-30}{5} = 0.5,$ and

  $at\,x = 41.5, z = \dfrac{41.5-30}{5} = 2.3$

  Hence using table, we find

  $$P(33 \leq X \leq 41) = P(0.5 \leq Z \leq 2.3)$$

  $$= P(0 \leq Z \leq 2.3) - P(0 \leq Z \leq 0.5)$$

  $$= 0.4893 - 0.1915 = 0.2978$$

- The discrete value 30, adjusted for continuity, becomes the interval $29.5\,to\,30.5,$ and the corresponding z values are:

  at $x = 29.5, z = \dfrac{29.5-30}{5} = -0.1,$ and

  at $x = 30.5, z = \dfrac{30.5-30}{5} = 0.1.$

  Hence using table, we obtain

  $$P(X = 30) = P(-0.1 \leq Z \leq 0.1)$$

  $$= 2(0.0398) = 0.0796.$$

Normal Approximation to the Poisson Distribution

The Poisson distribution $p(x;\mu)$ can also be approximated by the normal distribution, when $\mu \to \infty$. The probability for a Poisson r.v. X to take the value x is

$$P(X = x) = \frac{e^{-\mu}\mu^{x}}{x!}, for\, x = 0,1,2,\ldots,\infty.$$

And $E(X) = Var(X) = \mu.$

Let $Z = \dfrac{X-\mu}{\sqrt{\mu}}$ so that $E(Z) = 0\, and\, Var(Z) = 1.$

In order to show that as $\mu \to \infty$, the limiting distribution of Z is standard normal distribution, we use the moment generating function.

Now, the m.g.f. of Z is

$$M(t) = E\left(e^{tZ}\right) = E\left[e^{\frac{t(X-\mu)}{\sqrt{\mu}}}\right]$$

$$= e^{-t\sqrt{\mu}} E\left[e^{\frac{tx}{\sqrt{\mu}}}\right]$$

$$= e^{-\frac{t}{\sqrt{\mu}}} \sum_{x=0}^{\infty} e^{\frac{tx}{\sqrt{\mu}}} \cdot \frac{e^{-\mu}.\mu^{x}}{x!}$$

$$= e^{-t\sqrt{\mu}} \sum_{x=0}^{\infty} e^{\frac{tx}{\sqrt{\mu}}} \cdot \frac{e^{-\mu}.\mu^{x}}{x!}$$

$$= e^{-t\sqrt{\mu}}.e^{-\mu} \sum_{x=0}^{\infty} \frac{\left(\mu e^{-\frac{t}{\sqrt{\mu}}}\right)^{x}}{x!}$$

$$= e^{-t\sqrt{\mu}} . e^{-\mu} \left[1 + \mu e^{\frac{t}{\sqrt{\mu}}} + \frac{\left(\mu e^{-\frac{t}{\sqrt{\mu}}} \right)^2}{2!} + \ldots \right]$$

$$= e^{-t\sqrt{\mu}} . e^{-\mu} . e^{\mu . \frac{e^t}{\sqrt{\mu}}}$$

$$= e^{-t\sqrt{\mu}} . e^{\mu \left(\frac{e^t}{\sqrt{\mu - 1}} \right)}$$

Taking logs to the base e, we get

$$\log_e M(t) = -t\sqrt{\mu} + \mu \left(e^{-\frac{t}{\sqrt{\mu - 1}}} \right)$$

$$= -t\sqrt{\mu} + \mu \left[1 + \frac{t}{\sqrt{\mu}} + \frac{t^2}{2!\mu} + \frac{t^3}{3!\mu\sqrt{\mu}} + \ldots - 1 \right]$$

$$= -t\sqrt{\mu} + t\sqrt{\mu} + \frac{t^2}{2!} + \frac{t^3}{3!\sqrt{\mu}} + \frac{t^4}{4!\mu} + \ldots$$

$$= \frac{t^2}{2!} + \frac{t^3}{3!\sqrt{\mu}} + \frac{t^4}{4!\mu} + \ldots$$

Therefore $\lim_{\mu \to \infty} \log_e M(t) = \frac{t^2}{2!} or$

$$\lim_{\mu \to \infty} M(t) = e^{\frac{t^2}{2}}$$

Which is the m.g.f. of the standard normal distribution.

Hence we see that the Poisson distribution approaches the normal distribution as $\mu \to \infty$.

***Example*:** The number of calls received by an office switch board per hour follows a Poisson distribution with parameter 25. Find the probabilities that in one hour (a) there are between 23 and 26 calls (inclusive), (b) more than 30 calls, using the normal approximation to the Poisson Distribution.

Solution: Let X be the r.v. the number of calls received in one hour. Then $X \, is \, p(x;25)$. We require (a) $P(23 \le X \le 26)$ and (b) $P(X > 30)$.

Using the normal approximation, X is $N(25,25)$.

a. $P(23 \le X \le 26)$ becomes on continuous scale $P(22.5 \le X \le 26.5)$.

The z-values are: at $x = 22.5, z = \dfrac{22.5-25}{5} = -0.5,$ and

at $x = 26.5, z = \dfrac{26.5-25}{5} = 0.3$

$$P(22.5 \le X \le 26.5) = P(-0.5 \le Z \le 0.3)$$

$$= P(0 < Z < 0.5) + P(0 < Z < 0.3)$$

$$= 0.1915 + 0.1179 = 0.3094$$

b. $P(X > 30)$ becomes on continuous scale $P(X > 30.5)$

$$P(X > 30.5) = P\left(\frac{X-25}{5} > \frac{30.5-25}{5}\right) = P(Z > 1.1)$$

$$= 0.5 - P(0 < Z < 1.1)$$

$$= 0.5 - 0.3643 = 0.1357$$

Fitting a Normal Distribution

There are two possible situations to deal with.

a. Given an observed frequency distribution with classes, we need to calculate frequencies which we would expect for a normal distribution with the same mean and standard deviation as the given data.

b. Given only values of mean we need to determine the number of classes (and hence their width) for the presentation of the distribution.

Case (a) to Fit a Normal Distribution

To fit a normal distribution to an observed frequency distribution when neither the mean nor the variance is known, we proceed as below.

- We estimate the two parameters $\mu\, and\, \sigma^2$ by calculating $\bar{x}\, and\, s^2$ from the observed frequency distribution.
- We calculate the standard normal z- values corresponding to the upper class boundaries by subtracting the estimated mean from each upper class boundary and dividing by s. As the normal distribution is defined from $-\infty\, to\, \infty$, we therefore extend the lowest class backward to $-\infty$ and the last (highest) class forward to ∞.
- We find the cumulative probability $P(Z < z) = \varnothing(z)$ associated with each z-value, using the table of standard normal distribution.
- We then obtain the probability, $\hat{p}$ (or area) for each class by successive subtraction of $\varnothing(z)$.
- We finally get the expected (theoretical) frequencies by multiplying each of the class probabilities by the total frequency of the given distribution.

This procedure is also known as fitting a normal distribution to a given frequency distribution by area method.

Case (b) to Fit a Normal Distribution

When we are given only the values of the two parameters $\mu\, and\, \sigma^2$, we proceed as follows:

- We determine the practical range of the distribution as $\mu - 3\sigma\, to\, \mu + 3\sigma$, since most of the normal distributions are practically covered by a range of 6 standard deviations.
- We choose all the classes (between 6 and 15 classes) of a convenient within the range.
- We proceed as in Case (a) above to calculate the probabilities of the theoretical probability distribution. The expected

frequencies can be obtained when the total number of observation (total frequency) is available.

The following examples illustrate the procedure for fitting a normal distribution:

Example: Fit a normal distribution to the following frequency distribution of weights:

Weight	f_i
28–31	1
32–35	14
36–39	56
40–43	172
44–47	245
48–51	263
52–55	156
56-59	67
60–63	23
64–67	3

Solution: We first calculate the mean and standard deviation of this distribution to estimate the two parameters of the normal distribution.

Computation of mean and standard deviation.

Weight	f_i	x_i	$u_i = \frac{x_i - 49.5}{4}$	$f_i u_i$	$f_i u_i^2$
28-31	1	29.5	-5	-5	25
32-35	14	33.5	-4	-56	224
36-39	56	37.5	-3	-168	504
40-43	172	41.5	-2	-344	688
44-47	245	45.5	-1	-245	245
48-51	263	49.5	0	0	0

52-55	156	53.5	1	156	156
56-59	67	57.5	2	134	268
60-63	23	61.5	3	69	207
64-67	3	65.5	4	12	48
	1000			-447	2365

Now, $\overline{x} = a + \frac{\sum f_i u_i}{n} \times h$

$$= 49.5 + \frac{(-477)}{1000} \times 4$$

$$= 49.5 - 1.79 = 47.71 kg,$$

$$s = h \times \sqrt{\frac{\sum fu^2}{n} - \left(\frac{\sum fu}{n}\right)^2}$$

$$= 4 \times \sqrt{\frac{2365}{1000} - \left(\frac{-447}{1000}\right)^2}$$

$$= 4 \times \sqrt{2.365 - 0.1998}$$

$$= 4 \times 1.47 = 5.88 kg$$

The necessary calculations for the expected frequencies of the fitted distribution are shown below:

Weight	Upper class boundary	$z = \frac{u.c.b - \overline{x}}{s}$	$P(Z < z)\varnothing(z)$	Probability $(\hat{p})$	Expected Frequency
$-\infty - 27$	27.5	–3.44	0.0003	0.0003	3
$28 - 31$	31.5	–2.76	0.0029	0.0026	
$32 - 35$	35.5	–2.08	0.0188	0.0159	16
$36 - 39$	39.5	–1.40	0.0808	0.0620	62
$40 - 43$	43.5	–0.70	0.2389	0.1581	158
$44 - 47$	47.5	–0.03	0.4880	0.2491	249

48 – 51	51.5	0.64	0.7389	0.2509	251
52 – 55	55.5	1.32	0.9066	0.1677	168
56 – 59	59.5	2.01	0.9778	0.0712	71
60 – 63	63.5	2.68	0.9963	0.0185	18
64 – 67	67.5	3.37	0.9996	0.0033 }	4
68 – ∞			1.0000	0.0004 }	
	1000		–447	2365	1,000

SUMMARY AND DISCUSSION

In this chapter, we discussed about Normal distribution. The normal probability distribution, which is considered the corner stone of the modern statistical theory, was discovered by Abraham de Moivre (1667-175) as the limiting form of the binomial distribution by increasing n, the number of trials, to a very large number for a fixed value of p. But his paper remained unnoticed until 1924 when it was found in a library by Karl Pearson (1857-1936). The name of Pierre S. Laplace (1749-1827) is also associated with the derivation of the normal distribution.

The normal distribution is also called the Gaussian distribution in honor of the great German mathematician Carl F. Gauss (1777 - 1827), who also derived its equation mathematically as the probability distribution of the errors of measurements. It was Karl Pearson who in 1893 called it the normal distribution and is best known by this name today.

A normal distribution is defined by the p.d.f.

$$f(x) = \frac{1}{\sigma\sqrt{2\pi}} e^{-\left[\frac{(x-\mu)^2}{2\sigma^2}\right]}, \ for - \infty < x \langle \infty, and\ \sigma \rangle 0$$

Characteristics of Normal Distribution

- One characteristic of the normal distribution is that the mean and variance are independent. There is no association between them. Knowing one tells us nothing about the value of the other. This characteristic makes the normal distribution unusual.

- The normal distribution is symmetric.
- The normal distribution is unbounded both above and below. Hence the normal distribution is defined for all real numbers.
- Its importance stems from the Central limit theorem.
  - If what we observe in nature is the result of adding up lots of independent things, then the distribution of this sum tends to look normal the more things we add up.
  - As a result sample means tend to have a normal distribution when the sample size is big enough because in calculating a mean we add things up.
- Even if the response is a discrete random variable, like a count, a normal distribution may be an adequate approximation for its distribution if we're dealing with a large sample and the values we've obtained are far removed from any boundary conditions. On the other hand count data with lots of zero values cannot possibly have a normal distribution or be transformed to approximate normality.

Standardized Normal Distribution

A normal probability distribution depends on the values of the parameter $\mu \, and \, \sigma^2$ and the various possible values for these two parameters will result in an unlimited number of different normal distribution. The r.v. $Z = \frac{X - \mu}{\sigma}$, as we have seen, has zero mean and unit variance. Every normally distributed r.v. X with mean $\mu \, and \, variance = \sigma^2$ is therefore conveniently transformed into a new normal r.v Z with zero mean and unit variance by using the following expression

$$Z = \frac{X - \mu}{\sigma}$$

Then the p.d.f. of Z, denoted by $\phi(z)=\frac{1}{\sqrt{2\pi}}e^{-\frac{z^2}{2}},\ for -\infty < z < \infty.$

Properties of Normal Distribution

The main properties of the normal distribution are given below:

- The function $f(x)$ defining the normal distribution is a proper p.d.f. i.e. $f(x)\geq 0$ and the total area under the normal curve is unity.
- The mean and variance of the normal distribution are $\mu\ and\ \sigma^2$ respectively.
- The median and the mode of the normal distribution are each equal to $\mu,$ the mean of the distribution.
- The mean deviation of the normal distribution is approximately $\frac{4}{5}$ of its standard deviation.
- The normal curve has points of inflection which are equidistant from the mean.
- For the normal distribution, the odd order moments about the mean are all zero and the even order moments are given by
- $\mu_{2n}=(2n-1)(2n-3)\ldots 5.3.1\sigma^{2n}.$
- If X is $N(\mu,\sigma^2)$ and if $Y=a+bX,$ then Y is $N(a+b\mu,b^2\sigma^2).$
- The sum of independent normal variables is a normal variable. Stated differently, if X_1 is $N(\mu_1,\sigma_1^2)$ and X_2 is $N(\mu_2,\sigma_2^2),$ then for independent $X_1\ and\ X_2, X_1+X_2\ is$ $N(\mu_1+\mu_2,\sigma_1^2+\sigma_2^2).$
- No matter what the values of μ and σ are, areas under normal curve remain in certain fixed proportions within a specified number of standard deviations on either side of $\mu.$ For example, the interval

- $\mu \pm \sigma$ will always contain 68.26%
- $\mu \pm 2\sigma$ will always contain 95.44%
- $\mu \pm 3\sigma$ will always contain 99.73%

- The quartile deviation, Q is found as $\frac{1}{\sigma\sqrt{2\pi}}\int_{\mu-Q}^{\mu+Q} e^{-\frac{(x-\mu)^2}{2\sigma^2}}dx = \frac{1}{2}$
- The normal curve approaches, but never really touches, the horizontal axis on either side of the mean towards plus and minus infinity, that is the curve is asymptotic to the horizontal axis as $x \to \pm\infty$.

CHAPTER 9

REGRESSION AND CORRELATION

INTRODUCTION

The term regression was introduced by the English biometrician, Sir Francis Galton (1822-1911) to describe a phenomenon which he observed in analyzing the heights of children and their parents. He found that, though tall parents have tall children and short parents have short children, the average height of children tends to step back or to regress toward the average height of all men. This tendency toward the average height of all men was called a regression by Galton.

Today, the word regression is used in a quite different sense. It investigates the dependence of one variable, conventionally called the dependent variable, on one or more other variables, called independent variables, and provides an equation to be used for estimating or predicting the average value of the dependent variable from the unknown values of the independent variable. The dependent variable is assumed to be a random variable whereas the independent variables are assumed to have a fixed values, i.e. they are chosen non-randomly.

The relation between the expected value of the dependent variable and the independent variable is called a regression relation. When we study the independent variable it is called a regression relation. When we study the dependence of a variable on a single independent variable, it is called a dependence of a variable on a single independent variable, it is called a simple or two variable regression.

When the dependence of a variable on two or more than two independent variables is studied, it is called multiple regression. Furthermore, when the dependence is represented by a straight line equation, the regression is said to be linear, otherwise it is said to be curvilinear.

It is relevant to note that in regression study, a variable whose variation we try to explain is a dependent variable while an independent variable is a variable that is used to explain the variation in the dependent variable.

Some more terminology: The dependent variable is also called the regressand, the predictand, the response or the explained variable, whereas the independent or the non-random variable is also referred to as the regressor, the predictor, the regression variable or the explanatory variable.

DETERMINISTIC AND PROBABILISTIC RELATIONS OR MODELS

The relationship among variables may or may not be governed by an exact physical law. For convenience, let us consider a set of n pairs of observations (X_i, Y_i). If the relation between the variables is exactly linear, then the mathematical equation describing the linear relation is generally written as

$$Y_i = a + bX_i,$$

Where a is the value of $Y \, when \, X$ equal zero and is called the $Y - intercept$, and b indicates the change in Y for a one unit change in X and is called the slope of the line. Substituting a value for X in the equation, we can completely determine a unique value of Y. The linear relation in such a case is said to be a deterministic model. An important example of the deterministic model is the relationship between Celsius and Fahrenheit scales in the form of $F = 32 + \frac{9}{5}C.$ Another example is the area of a circle expressed by the relation, area $= \pi r^2$. Such relations cannot be studied by regression.

In contrast to the above, the linear relationship in some situations is not exact. For example, we cannot precisely determine a person's weight from his height as the relationship between them is not expected to follow an exact linear form. The weights for given values of age are reasonably assumed to include measurement of random errors. The deterministic relation in such cases is then modified to allow for the in exact relationship between the variables and we get what is called a non-deterministic or probabilistic model as

$$Y_i = a + bX_i + e_i, \quad (i = 1, 2, \ldots, n)$$

Where e_i's are the unknown random errors.

Simple Linear Regression Model

We assume that the linear relationship between the dependent variable Y_i and the value X_i of the regressor X is

$$Y_i = \alpha + \beta X_i + \varepsilon_i$$

Where the X_i's are fixed or predetermined values,

The Y_i's are observations randomly drawn from a population,

The ε_i's are errors components or random deviations,

α and β are population parameters, α is the intercept and the slope β is called regression coefficient, which may be positive or negative depending upon the direction of the relationship between X and Y.

Furthermore, we assume that

- $E(\varepsilon_i) = 0$, i.e. the expected value of error term is zero, it implies that the expected value of Y is related to X in the population by a straight line;
- $Var(\varepsilon_i) = E(\varepsilon_i^2) = \sigma^2$ for all I, i.e. the variance of error term is constant. It means that the distribution of error has the same variance for all values of X. (Homoscedasticity assumption);

- $E(\varepsilon_i, \varepsilon_j) = 0$ for all $i \neq j$, i.e. error terms are independent of each other (assumption of no serial or auto correlation between $\varepsilon's$)
- $E(X, \varepsilon_i) = 0$, i.e. X and ε are also independent of each other
- $\varepsilon_i's$ are normally distributed with a mean of zero and a constant variance σ^2. This implies that Y values are also normally distributed. The distributions of $Y \, and \, \varepsilon$ are identical except that they have different means. This assumption is required for estimation and testing of hypothesis on linear regression.

According to this population regression model, each Y_i is an observation from a normal distribution with mean $= \alpha + \beta x$ and variance $= \sigma^2$. Thus the relation may be expressed alternatively as $E(Y) = \alpha + \beta X$, which implies that the expected value of Y is linearly related to X and the observed value Y deviates from the line $E(Y) = \alpha + \beta X$ by a random components $\varepsilon, i.e.\ \varepsilon_i = Y_i - (\alpha + \beta X_i)$. The following graph illustrates the assumed line, giving $E(Y)$ for the given values of X.

But in practice, we have a sample from some population, therefore we desire to estimate the population regression line from the sample data. Then the basic relation in terms of sample data may be written as $Y_i = a + bX_i + e_i$

Where a, b and e_i are the estimates of $\alpha, \beta \, and \, \varepsilon_i$. The estimated regression is generally written as $\widehat{Y}_i = a + b X_i$.

Many possible regression lines could be fitted to the sample data, but we choose that particular line which best fits that data. The best regression line is obtained by estimating the regression parameters by the most commonly used method of least square which we describe in the following subsections.

Least Square

The method of least square that gets its name from the minimization of a sum of squared deviations, is attributed to Karl F. Gauss (1777-1855). Some people believe that the method was discovered at the same time by Adrien M. Legendre (1752-1833), S. Laplace (1749-1827) and others. Markov's name is also mentioned in connection with its further development. In recent years, efforts have been made to find better methods of fitting but the least squares methods remains dominant and is used as one of the important methods of estimating the population parameters.

Least Square Estimates in Simple Linear Regression

Let there be a set of observation $\{(X_i, Y_i), i = 1, 2, \ldots, n\}$, where Y_i are the values of Y randomly drawn from a population and X_i are fixed values. Then the observed Y_i may be expressed in a linear form of the population parameters as $Y_i = \alpha + \hat{a}\ X_i + \mathring{a}_i$, or in terms of sample data as $Y_i = a + bX_i + e_i$, where a and b are the least squares estimates of $\alpha\ and\ \beta, e_i$ commonly called residual, is the deviation of the observed Y_i from its estimate provided by $\widehat{Y}_i = a + bX_i$.

According to the principle of least squares, we determine those values of a and b which will minimize the sum of squares of the residuals. In other words, the best regression line is one which minimize the sum of the squares of the vertical deviations between the observed values Y_i, and the corresponding values predicated by the regression model, i.e. $\widehat{Y}_i = a + bX_i$. That is, the least square line minimizes

$$S(a,b) = \sum_{i=1}^{n} e_i^2 = \sum_{i=1}^{n} \left(Y_i - \hat{Y}\right)^2$$

$$= \sum \left(Y_i - a - bX_i\right)^2$$

As a and b, the two quantities that determine the line, vary, $S(a,b)$ will vary too. We therefore consider $S(a,b)$ as a function of a and b, and we wish to determine at what values of a and b, it will be minimum.

Minimizing $S(a,b)$, we need to set its partial derivative *w.r.t* a and b equal to zero. Therefore

$$\frac{\partial S(a,b)}{\partial a} = 2\Sigma(Y_i - a - bX_i)(-1) = 0,$$

and

$$\frac{\partial S(a,b)}{\partial b} = 2\Sigma(Y_i - a - bX_i)(-X_i) = 0,$$

Simplifying, we obtain the following two equations, called the normal equations (the word normal is used here in the sense of regular or standard).

$$\Sigma Y_i = na + b\Sigma X_i \quad (1)$$

$$\Sigma X_i Y_i = a\Sigma X_i + b\Sigma X_i^2 \quad (2)$$

These two normal equations are solved simultaneously for the values of a and b either by direct elimination or by using determinants.

- Direct Elimination: Multiplying the first equation by ΣX_i and the second equation by n, we get

$$\Sigma X \Sigma Y = na\Sigma X + b(\Sigma X)^2 \quad (3)$$

$$n\Sigma XY = na\Sigma X + nb\Sigma X^2 \quad (4)$$

Subtracting, we get

$$n\Sigma XY - \Sigma X \Sigma Y = b\left[n\Sigma X^2 - (\Sigma X)^2\right]$$

Therefore $$b = \frac{n\Sigma XY - \Sigma X \Sigma Y}{n\Sigma X^2 - (\Sigma X)^2} = \frac{\Sigma(X - \bar{X})(Y - \bar{Y})}{\Sigma(X - \bar{X})^2}$$

Which is the least square estimate of the regression coefficient β.

Similarly, we get

$$a = \frac{\sum X^2 \sum Y - \sum X \sum XY}{n\sum X^2 - (\sum X)^2}$$

As the least square estimate of α.

It is very important to note that, when both $X \, and \, Y$ are observed at random, i.e. the sample values are from a bivariate population, there are two regression equations, each obtained by choosing that variable as dependent whose average value is to be estimated and treating the other variable as independent. In case of a single random variable, the single regression equation is used to estimate the values of either the dependent or the independent variable. In case of two regression lines, it is customary to denote the regression coefficient of Y on X and $X \, on \, Y$ by $b_{yx} \, and \, b_{xy}$ respectively.

***Example*:** Compute the least square regression equation of for the following data. What is the regression coefficient and what does it mean?

X	5	6	8	10	12	13	15	16	17
Y	16	19	23	28	36	41	44	45	50

***Solution*:** The estimated regression line of $Y \, on \, X$ is

$$\hat{Y} = a + bX$$

And the two normal equations are

$$\sum Y = na + b\sum X$$

$$\sum XY = a\sum X + b\sum X^2.$$

$$\bar{X} = \frac{\sum X}{n} = \frac{102}{9} = 11.33, \bar{Y} = \frac{\sum Y}{n} = \frac{302}{9} = 33.56,$$

$$b_{yx} = \frac{n\sum XY - \sum X \sum Y}{n\sum X^2 - (\sum X)^2}$$

$$b_{yx} = \frac{9(3853) - (102)(302)}{9(1308) - (102)^2}$$

To compute the necessary summations, we arrange the computations in the table below:

X	Y	XY	X^2
5	16	80	25
6	19	114	36
8	23	184	64
10	28	280	100
12	36	432	144
13	41	533	169
15	44	660	225
16	45	720	256
17	50	850	289

$$b_{yx} = \frac{34677-30804}{11772-10404} = \frac{3873}{1368} = 2.831,$$ and

$$a = \overline{Y} - b\overline{X} = 33.56 - (2.831)(11.33) = 1.47$$

Hence the desired estimated regression line of $Y \, on \, X$ is $\hat{Y} = 1.47 + 2.831X.$

The estimated regression coefficient, $b = 2.831,$ which indicates that the values of Y increase by 2.831 units for a unit increase in $X.$

***Example*:** In an experiment to measure the stiffness of a spring, the length of the spring under different loads was measured as follows:

$X = Loads(lb)$	3	5	6	9	10	12	15	20	22	28
$Y = length(in)$	10	12	15	18	20	22	27	30	32	34

Find the regression equation appropriate for predicting

- The length, given the weight on the spring;
- The weight, given the length of the spring.

***Solution*:** The data come from a bivariate population, i.e. both $X \text{ and } Y$ are random, therefore there are two regression lines. To find the regression equation for predicting length (Y), we take Y as dependent variable and treat X as independent variable (i.e. non-random). For the second regression, the choice of the variable is reserved.

The computations needed for the regression lines are given in the following table:

X	Y	X^2	Y^2	XY
3	10	9	100	30
5	12	25	144	60
6	15	36	225	90
9	18	81	324	162
10	20	100	400	200
12	22	144	484	264
15	27	225	729	405
20	30	400	900	600
22	32	484	1024	704
28	34	784	1156	932
130	220	2288	5486	3467

- The estimated regression equation appropriate for predicting the length, Y, given the weight X is

$$\hat{Y} = a_0 + b_{yx}X,$$

where

$$b_{yx} = \frac{n\sum XY - \sum X \sum Y}{n\sum X^2 - \left(\sum X\right)^2} = \frac{(10)(3467) - (130)(220)}{(10)(2288) - (130)^2}$$

$$= \frac{6070}{5980} = 1.02,$$ and

$$a_0 = \overline{Y} - b_{yx}\overline{X} = 22 - (1.02)(13) = 8.74$$

Hence the desired estimated regression equation is

$$\hat{Y} = 8.74 + 1.02X$$

- The estimated regression equation appropriate for predicting the weight, X, given the length is

$$\hat{X} = a_1 + b_{xy}Y,$$

where

$$b_{xy} = \frac{n\Sigma XY - (\Sigma X)(\Sigma Y)}{n\Sigma Y^2 - (\Sigma Y)^2} = \frac{(10)(3467) - (130)(220)}{(10)(5486) - (220)^2}$$

$$= \frac{6070}{6460} = 0.94,$$ and

$$a_1 = \overline{X} - b_{xy}\overline{Y} = 13 - (0.94)(22) = -7.68$$.

Hence $\hat{X} = 0.94Y - 7.68$ is the estimated regression equation appropriate for predicting the weight (X), iven the length (Y).

Properties of the least square

The least square linear regression line has the following properties.

- The least squares regression line always goes through the point $(\overline{X}, \overline{Y})$, the means of the data.
- The sum of the deviations of the observed values of Y_i from the least square regression line is always equal to zero, i.e. $\Sigma(Y_i - \hat{Y}) = 0$

- The sum of the squares of the deviations of the observed values from the least square regression line is a minimum, i.e. $\Sigma\left(Y_i - \widehat{Y}_i\right)^2 = minimum$
- The least square regression line obtained from a random sample is the line of best fit because a and b are the unbiased estimates of the parameters α *and* β.

CORRELATION

Correlation, like covariance, is a measure of the degree to which any two variables vary together. In other words, two variables are said to be correlated if they tend to increase (or decrease) together, the correlation is said to be direct or positive, and e.g. the length of an iron bar will increase as the temperature increases. If one variable tends to increase as the other variable decreases, the correlation is said to be negative or inverse, e.g. the volume of gas will decrease as the pressure increase. It is worth remarking that in correlation, we assess the strength of the relationship (or interdependence) between two variables; both the variables are random variables, and they are treated symmetrically, i.e. there is no distinction between dependent and independent variable. In regression by contrast, we are interested in determining the dependence of one variable that is random, upon the other variable that is non-random or fixed, and in predicting the average value of the dependent variable by using the known values of the other variable.

Pearson Product Moment Correlation Coefficient

A numerical measure of strength in the linear relationship between any two variables is called the Pearson's product moment correlation coefficient or sometimes, the coefficient of simple correlation or total correlation. The sample linear correlation coefficient for n pairs of observations (X_i, Y_i) usually denoted by the letter r is defined by

$$r = \frac{\Sigma\left(X - \overline{X}\right)\left(Y - \overline{Y}\right)}{\sqrt{\Sigma\left(X - \overline{X}\right)^2 \Sigma\left(Y - Y\right)^2}}$$

The population correlation coefficient for a bivariate distribution, denoted by ρ, has already been defined as

$$\rho = \frac{Cov(X,Y)}{\sqrt{Var(X)Var(Y)}}$$

For computational purposes, we have an alternative form of r as

$$r = \frac{\sum XY - \frac{(\sum X)(\sum Y)}{n}}{\sqrt{\left[\sum X^2 - \frac{(\sum X)^2}{n}\right]\left[\sum Y^2 - \frac{(\sum Y)^2}{n}\right]}}$$

This is more convenient and useful form, especially when $\bar{X}\ and\ \bar{Y}$ are not integers. The coefficient of correlation r is a pure number (i.e. independent of the units in which the variable are measured) and it assumes values that can range from +1 for perfect positive linear relationship, to -1 for perfect negative linear relationship with the intermediate value of zero indicating no linear relationship between $X\ and\ Y$. The sign of r indicates the direction of the relationship or correlation.

It is important to note that $r = 0$ does not mean that there is no relationship at all. For example, if all the observed values lie exactly on a circle, there is a perfect non-linear relationship between the variables but r will have a value of zero as r only measure the linear correlation.

The linear correlation coefficient, is also the square root of the linear coefficient of determination, r^2.

Example: Calculate the product moment coefficient of correlation between $X\ and\ Y$ from the following data:

X	1	2	3	4	5
Y	2	5	3	8	7

Solution: The calculation needed to compute r are given below:

X	Y	X^2	Y^2	XY
1	2	1	4	2
2	5	4	25	10
3	3	9	9	9
4	8	16	64	32
5	7	25	49	35
15	25	55	151	88

$$r = \frac{\sum XY - \frac{(\sum X)(\sum Y)}{n}}{\sqrt{\left[\sum X^2 - \frac{(\sum X)^2}{n}\right]\left[\sum Y^2 - \frac{(\sum Y)^2}{n}\right]}}$$

$$= \frac{88 - \frac{(15)(25)}{5}}{\sqrt{\left[55 - \frac{(15)^2}{5}\right]\left[151 - \frac{(25)^2}{5}\right]}}$$

$$= \frac{13}{\sqrt{10 \times 26}} = 0.8$$

Properties:

The sample correlation coefficient r has the following properties:

- The correlation coefficient r is symmetrical with respect to the variable $X\ and\ Y$, i.e. $r_{XY} = r_{YX}$.
- The correlation coefficient lies between -1 an $+1$, i.e. $-1 \leq r \leq +1$.
- The correlation coefficient is independent of the origin and scale.

SUMMARY AND DISCUSSION

In this chapter we discussed about two important topics of statistics, regression and correlation. Both topics play a very important role in Statistical theory and also in real life problems.

It investigates the dependence of one variable, conventionally called the dependent variable, on one or more other variables, called independent variables, and provides an equation to be used for estimating or predicting the average value of the dependent variable from the unknown values of the independent variable. The dependent variable is assumed to be a random variable whereas the independent variables are assumed to have fixed values, i.e. they are chosen non-randomly.

The relation between the expected value of the dependent variable and the independent variable is called a regression relation. When we study the independent variable, it is called a regression relation. When we study the dependence of a variable on a single independent variable, it is called a dependence of a variable on a single independent variable, it is called a simple or two variable regression.

When the dependence of a variable on two or more than two independent variables is studied, it is called multiple regression. Furthermore, when the dependence is represented by a straight line equation, the regression is said to be linear, otherwise it is said to be curvilinear.

Simple Linear Regression Model

We assume that the linear relationship between the dependent variable Y_i and the value X_i of the regressor X is

$Y_i = \alpha + \beta X_i + \varepsilon_i$

where the X_i's are fixed or predetermined values,

the Y_i's are observations randomly drawn from a population,

the ε_i's are errors components or random deviations,

α and β are population parameters; α is the intercept and the slope

β is called regression coefficient, which may be positive or negative depending upon the direction of the relationship between $X \ and \ Y$.

Normal Equations

$$\sum Y_i = na + b\sum X_i$$

$$\sum X_i Y_i = a\sum X_i + b\sum X_i^2$$

$$b = \frac{n\sum XY - \sum X \sum Y}{n\sum X^2 - (\sum X)^2} = \frac{\Sigma(X-\bar{X})(Y-\bar{Y})}{\Sigma(X-\bar{X})^2}$$

Which is the least square estimate of the regression coefficient β.

Similarly, we get

$$a = \frac{\sum X^2 \sum Y - \sum X \sum XY}{n\sum X^2 - (\sum X)^2}$$

As the least square estimate of α.

Properties of the Least Square

The least square linear regression line has the following properties.

- The least squares regression line always goes through the point $(\bar{X}, \bar{Y})$, the means of the data.
- The sum of the deviations of the observed values of Y_i from the least square regression line is always equal to zero, i.e. $\Sigma(Y_i - \hat{Y}) = 0$
- The sum of the squares of the deviations of the observed values from the least square regression line is a minimum, i.e. $\Sigma(Y_i - \hat{Y}_i)^2 = minimum$
- The least square regression line obtained from a random sample is the line of best fit because a and b are the unbiased estimates of the parameters $\alpha \ and \ \beta$.

Pearson Product Moment Correlation Coefficient

A numerical measure of strength in the linear relationship between any two variables is called the Pearson's product moment correlation coefficient or sometimes, the coefficient of simple correlation or total correlation. The sample linear correlation coefficient for n pairs of observations (X_i, Y_i) usually denoted by the letter r is defined by

$$r = \frac{\Sigma\left(X - \overline{X}\right)\left(Y - \overline{Y}\right)}{\sqrt{\Sigma\left(X - \overline{X}\right)^2 \Sigma\left(Y - Y\right)^2}}$$

The population correlation coefficient for a bivariate distribution, denoted by ρ, has already been defined as $\rho = \frac{Cov(X,Y)}{\sqrt{Var(X)Var(Y)}}$

For computational purposes, we have an alternative form of r as

$$r = \frac{\Sigma XY - \frac{(\Sigma X)(\Sigma Y)}{n}}{\sqrt{\left[\Sigma X^2 - \frac{(\Sigma X)^2}{n}\right]\left[\Sigma Y^2 - \frac{(\Sigma Y)^2}{n}\right]}}$$

***Properties*:**

The sample correlation coefficient r has the following properties:

- The correlation coefficient r is symmetrical with respect to the variable $X \, and \, Y$, i.e. $r_{XY} = r_{YX}$;
- The correlation coefficient lies between -1 an $+1$, i.e. $-1 \leq r \leq +1$;
- The correlation coefficient is independent of the origin and scale.

REFERENCES

1. Ronald E. Walpole, Raymond H. Myers, Sharon L. Myers and Keying Ye. *Probability and Statistics for Engineers and Scientists*, New York: Prentice Hall, 2011.
2. Dimitri P. Bertsekas, John N. Tsitsiklis, *Lecture Notes on Introduction to Probability*, Massachusetts Institute of Technology Cambridge, Massachusetts.
3. F. Catherine, E. Merran, H. Nicholas and P. Brian. *Statistical Distribution*, New Jersey: John Wiley and Sons, 2009.
4. Robert B. Ash, *Basic Probability Theory*, New York: Dover Publications, 2008.
5. https://www.britannica.com/topic/probability-theory
6. http://www.math.uiuc.edu/~r-ash/BPT.html
7. http://www.probabilitytheory.info/
8. https://en.wikipedia.org/wiki/Probability_theory

INDEX